About Island Press

Since 1984, the nonprofit organization Island Press has been stimulating, shaping, and communicating ideas that are essential for solving environmental problems worldwide. With more than 1,000 titles in print and some 30 new releases each year, we are the nation's leading publisher on environmental issues. We identify innovative thinkers and emerging trends in the environmental field. We work with world-renowned experts and authors to develop cross-disciplinary solutions to environmental challenges.

Island Press designs and executes educational campaigns, in conjunction with our authors, to communicate their critical messages in print, in person, and online using the latest technologies, innovative programs, and the media. Our goal is to reach targeted audiences—scientists, policy makers, environmental advocates, urban planners, the media, and concerned citizens—with information that can be used to create the framework for long-term ecological health and human well-being.

Island Press gratefully acknowledges major support from The Bobolink Foundation, Caldera Foundation, The Curtis and Edith Munson Foundation, The Forrest C. and Frances H. Lattner Foundation, The JPB Foundation, The Kresge Foundation, The Summit Charitable Foundation, Inc., and many other generous organizations and individuals.

The opinions expressed in this book are those of the author(s) and do not necessarily reflect the views of our supporters.

The Hype About Hydrogen

REVISED & UPDATED EDITION

The Hype
About Hydrogen

REVISED & UPDATED EDITION

FALSE PROMISES AND REAL SOLUTIONS IN THE
RACE TO SAVE THE CLIMATE

Joseph J. Romm

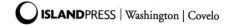
ISLANDPRESS | Washington | Covelo

Library of Congress Control Number: 2024946710

All Island Press books are printed on environmentally responsible materials.

Manufactured in the United States of America
10 9 8 7 6 5 4 3 2 1

Keywords: blue hydrogen, carbon capture, carbon capture and storage (CCS), carbon dioxide, carbon emissions, climate change, Department of Energy (DOE), direct air capture (DAC), e-fuels, electric cars, fossil fuel industry, fossil fuels, fuel cell, fusion, green hydrogen, greenhouse gas, hydrogen, hydrogen cars, Intergovernmental Panel on Climate Change (IPCC), net zero, nuclear, renewable energy, solar power, wind powercience education, science skeptics, science-society interface, scientific impact

To Antonia

Contents

Once I got home, I sulked for a while.
All my brilliant plans foiled by thermodynamics.
Damn you, Entropy!

—Andy Weir, *The Martian*

Foreword

HYDROGEN, DIRECT AIR CAPTURE, AND GEOENGINEERING are the worst of the almost-good ideas for addressing climate change. They sound like they could be good, but they aren't.

Hydrogen is politically appealing because it makes us feel like we can continue doing things just as we did in the past. Hydrogen has a strong and well-financed lobby group—specifically the oil and gas industry because it allows fossil fuel companies to continue business as usual while selling the false belief that hydrogen will come along as a drop-in substitute for how we do things, just in time to save the day.

Hydrogen also appeals to federal research and development programs because it can appear apolitical—a solution that, if it were true, steps on no one's toes. Engineers love working on it because the problems are challenging, and program managers like funding it because they get patted on the back by the same lobbyists pushing governments toward this distraction.

We will need some hydrogen in the future but primarily as a chemical feedstock rather than an energy solution. The question is how much.

Even the justification that we need huge amounts of it for making ammonia fertilizers should be called into question given what we know about the negative downstream consequences for soils and waterways of too much fertilizer.

The International Energy Agency (IEA) was born of the 1973 oil embargo and was established to ensure supply chains of fossil fuels for the Organisation for Economic Co-operation and Development. This cabal-like origin is why the IEA took so long to acknowledge (nonfossil) climate solutions. For years the IEA modeled ridiculous estimates that more than 20% of the world's electricity would be used for making hydrogen and that hydrogen would be a large fraction of the world's energy supply. Unfortunately, many of the world's governments embraced this convenient story and discredited the scientists and engineers advocating renewable electricity.

Part of the problem is the history of hydrogen in narratives about energy security. Both Japan and Germany pursued hydrogen as part of a plan to turn coal into high-density liquid fuels during World War II. Oil was military power, and neither had substantial domestic oil supplies. Without oil pipelines, Germany and Japan couldn't sustain the war. Hydrogen was meant to be part of the fix for that and thereby provide energy security.

That energy security story has changed to one of having an importable fuel that could come from multiple sources. The auto industries of both countries bet heavily on hydrogen, while China and Tesla went after electrification. They are still struggling to catch up. Betting on gaseous fuels is a costly mistake for a lot of countries, including the gas-led recovery in Australia and the hundreds of billions of dollars available for hydrogen energy in the Inflation Reduction Act, including the most dubious kind, hydrogen from natural gas with carbon capture and storage.

In the abstract, hydrogen sounds like a sort-of-good idea. Its energy per unit of weight is more than twice that of diesel fuel. But it's such a diffuse gas that the energy it carries per unit volume, even under impossible pressures or cooled to ridiculous temperatures, is one third that of gasoline or diesel—before you consider the weight of the storage vessels. It sounds like a substitute for "natural" gas, except you can't use existing gas infrastructure to pump it around because the molecule is too small. It sounds serious, as if you could run an industry on it, except you will have to change all of an industry's processes to accommodate it.

Hydrogen barely needed more problems to highlight its impracticality, but recent studies show that the global warming potential over twenty years—and people advocating hydrogen do so purportedly to solve global warming—is thirty-five times that of carbon dioxide! Given that 10%–20% of it leaks for liquid hydrogen storage and transport, this is a manifest disaster and means it probably will never be a significant contributor to desperately needed climate solutions.

Hydrogen is flammable. Hydrogen is explosive. Hydrogen is incredibly leaky. Hydrogen leaks speed up global warming. Hydrogen is a very inefficient way to convert renewable or nuclear electricity into useful energy. All of this results in the terminal problem of hydrogen: It's going to be expensive and challenging to work with. If you can do it with hydrogen, you can probably do it for one fifth as much with electricity.

And, of course, none of what I'm saying here is new. Joseph Romm should be applauded for pointing all this out (and more) for more than twenty years—twenty years and billions, perhaps trillions of dollars we have wasted on sort-of-good ideas that we could have spent on actually decarbonizing things and moving the planet to a sustainable and largely all-electric future.

This is a must-read book for anyone who thinks a hydrogen economy is a better investment than an electrified one. So let's stop believing the hype about hydrogen once and for all and start embracing the electrifying project of making this planet great again.

—Saul Griffith, Ph.D.
Author of *Electrify: An Optimist's Playbook for Our Clean Energy Future*

INTRODUCTION

In his January 2003 State of the Union address, George W. Bush announced a major initiative laying out his vision of a hydrogen economy: "Tonight, I'm proposing $1.2 billion in research funding so that America can lead the world in developing clean, hydrogen-powered automobiles. A single chemical reaction between hydrogen and oxygen generates energy, which can be used to power a car—producing only water, not exhaust fumes. With a new national commitment, our scientists and engineers will overcome obstacles to taking these cars from laboratory to showroom, so that the first car driven by a child born today could be powered by hydrogen, and pollution-free."

That speech was the reason I wrote the first edition of this book—which was published in 2004—to answer the question of whether such a high-tech Eden was hype that was too good to be true or whether it would happen in your lifetime.

But although I supported increasing the budget for hydrogen in the 1990s when I helped oversee the program at the Department of Energy, I slowly changed my mind after talking to dozens of experts, reading dozens of studies, and doing my own analysis for that book. Hydrogen simply had too many fatal limitations.

So, two decades ago, I predicted that this Eden was a myth and that hydrogen vehicles could not compete with plug-in vehicles. And indeed, today, electric vehicles outsell hydrogen ones 1,000-to-1.

The dream of a hydrogen economy has gone nowhere, particularly when you compare it with the electrification economy, which is widely understood to be the most efficient, affordable, and scalable set of solutions to climate change. Today, as the International Energy Agency (IEA) explained in 2023, "Hydrogen is mainly used in the refining and chemical sectors and produced using fossil fuels such as coal and natural gas, and thus responsible for significant annual CO_2 emissions."[1] While hydrogen demand keeps growing, it "remains concentrated in traditional applications." The traditional application of hydrogen is as a chemical feedstock. For instance, hydrogen is needed to make ammonia fertilizer because ammonia is NH_3, one nitrogen atom and three hydrogen atoms.

But the dream of a hydrogen economy is not built around using pollution-free hydrogen as a feedstock, which would reduce global greenhouse gas (GHG) emissions by no more than 2%. It's built on hydrogen as an energy carrier, much as electricity is an energy carrier, in the hopes of having ten times that impact (or more) on GHG emissions. In the energy delivery system, energy carriers lie between primary sources such as coal or wind and end-use applications such as vehicles, power plants, and home heating. They are "transmitters of energy." Like electricity, hydrogen is not a readily accessible energy source like coal or wind. Hydrogen is bound up tightly in molecules such as water (H_2O) and natural gas, which is mostly methane (CH_4), so it is expensive and energy-intensive to extract and purify.

Australian electrochemist John Bockris, who is credited with originating the phrase "hydrogen economy" in the early 1970s, wrote in 2002, "Boiled down to its minimalist description, the 'Hydrogen Economy' means that hydrogen would be used to transport energy from renewables (at nuclear or solar sources) over large distances; and to store it (for supply to cities) in large amounts."[2] Our cars, homes, and industries would be powered not by pollution-generating fossil fuels but by hydrogen from pollution-free sources.

In recent years, climate experts have focused on the goal of "green" hydrogen, or hydrogen made by electrolyzing water with renewable power. This became the key goal because, for hydrogen to be of any value in the fight against climate change, it must be made in a way that releases little, if any, GHGs, of which carbon dioxide is the most important.

Unfortunately, the cost of producing green hydrogen is still high, even though the idea seems obvious and tidy: Use electricity from a wind turbine to run current through water, "electrolyzing" it to split off the hydrogen molecules. But nothing involving hydrogen turns out to be easy. So predictions have blown up.

Electrolyzer prices jumped 50% between 2022 and 2023, according to S&P Global Commodities, because "electrolyzer projects tend to be highly complex, bespoke, and are proving far harder to construct than initially anticipated."[3] A 2023 Boston Consulting Group analysis noted electrolyzers used to make green hydrogen production "have a cost-overrun potential exceeding 500%."[4]

A September 2024 BloombergNEF study forecast electrolysis system capital costs falling by only "about one half from today to 2050 in China, Europe and the US assuming continued government support and free trade." The market research firm explains that this forecast "is about three times as high as what we anticipated in our 2022 analysis."[5]

Claims that carbon-free hydrogen energy will soon be a marketplace winner assume that costs of key equipment must steadily drop over time—and will do so faster than the competition. But even though the price of renewables has dropped sharply in the last two decades, the process of making hydrogen from renewables, storing it, and transporting it over large distances remains expensive and inefficient. That's why 98% of hydrogen today is still produced from fossil fuels—hydrocarbons—in processes that generate large quantities of CO_2.

After decades of effort and billions in spending by governments, car companies, and others, "the uptake of hydrogen in new applications in heavy industry, transport, the production of hydrogen-based fuels or electricity generation and storage . . . remains minimal, accounting for less than 0.1% of global demand," as the IEA reported in September 2023.[6] All the major novel applications of hydrogen—where it's used as an energy carrier—have failed in the market because hydrogen is a lousy energy carrier. If it had been a good energy carrier, we probably would have started using it as one a long time ago.

The central focus of this book is to explain why hydrogen is such a poor energy carrier—especially compared with electricity—and, therefore, why its role as a climate solution will be limited before 2050, if ever.

Indeed, for the foreseeable future, hydrogen should be considered a climate problem, not a climate solution. The world has not even started to address the CO_2 emissions from hydrogen's traditional use as a chemical feedstock for petrochemicals and fertilizers. That 100 million tons a year of hydrogen is responsible for about 2% of global CO_2 emissions. Simply replacing that dirty feedstock hydrogen with green hydrogen would require generating as much renewable electricity as the United States produces each year from all sources, renewable and nonrenewable. Achieving that goal by 2050—without significantly increasing hydrogen emissions—is implausible. The chances we would simultaneously be able to turn hydrogen into a successful energy carrier are vanishingly low.

The environmental paradise of a clean hydrogen economy rests not only on green hydrogen itself but also on a fuel cell for converting it into useful energy without generating pollution. A fuel cell is a small, modular electrochemical device, similar to a battery, but which can be continuously fueled. For most purposes, think of a fuel cell as a black box that takes in hydrogen and oxygen and puts out water plus electricity and heat but no pollution.

But just as hydrogen has not become a competitive energy carrier, fuel cells have not become a competitive conversion device even though they were invented nearly two centuries ago. In 1839 London, William Grove was experimenting with electrolysis, decomposing water into oxygen and hydrogen by using electricity. He figured out how to reverse the process to generate electricity by combining oxygen and hydrogen. Grove then quickly built the first fuel cell, which he called a "gaseous voltaic battery," or gas battery.

Yet here is the astonishing thing, as reported by Canary Media in June 2023: "Since the invention of the fuel cell in 1839, no company has been able to earn a profit from this energy-generating technology. Until the historic moment in which we now find ourselves."[7] In 2022, the article explains, a German manufacturer of methanol-powered fuel cells announced $1.8 million in profit.

But that company had *lost* a combined $22 million in the previous six years, and it was focused on niche applications: portable and remote use. We may have a long wait before a company making a fuel cell that could actually enable a broader hydrogen economy no longer has total losses that exceed total profits. The other four publicly traded fuel cell companies have combined cumulative losses of about $10 billion. And each of them saw their highest losses ever in 2022.

So, 180 years after fuel cells were invented, the industry has failed to live up to its promises. After decades of effort by scientists and entrepreneurs and billions in subsidies, hydrogen is not a successful energy carrier, and fuel cells are not a successful energy conversion device for hydrogen. At this point, we have no reason to believe they will be even a modest climate solution by 2050.

With so many obstacles remaining, major government policy should not be based on any claim of inevitability for green hydrogen as a major energy carrier. Government investments can and should be limited and slow. We must minimize both the cost of misallocating vast amounts of renewables and money and the likelihood of massive, stranded assets built for a hydrogen economy that never materializes.

You wouldn't use hydrogen to decarbonize either a sector or a niche application unless every other plausible solution had failed. And if you can do something with electricity, hydrogen will never be able to compete seriously. That was clear two decades ago, as I spelled out in the 2005 paperback edition of this book.

Hydrogen and fuel cells don't merely have to overcome their many limitations to succeed in the marketplace. They must also surpass and defeat their main competition—electricity and electrification technologies—which have been improving in cost and performance much faster for decades. Even though the world is diverting tens of billions of dollars—and potentially hundreds of billions—to hydrogen and enabling technologies, its fatal flaws today are the same as in the early 2000s.

First and foremost, hydrogen vehicles are an incredibly inefficient way to use carbon-free energy. Converting renewable energy to green hydrogen with an electrolyzer—then transporting, storing, and pumping it into a vehicle's onboard fuel cell just to regenerate electricity to turn an electric motor—means throwing away as much as 80% of the original renewables. And it relies on expensive electrolyzers and fuel cells.

That complex process can't compete with simply using the original renewable electricity to charge the battery on an electric vehicle and then running the electric motor off the battery, which loses less than 20% of the renewables. So you can go four times farther on the same renewable power with an electric vehicle than a hydrogen car—at a far lower cost.

The hydrogen scheme is as if you wanted someone to ship you bottled water—but to do that, they used the water to make champagne, shipped the champagne to your destination under expensive, carefully controlled conditions, and then distilled it back into water before delivering it to you.

For the same reason, hydrogen is a lousy way to try to store solar or wind power for use when the sun isn't shining or the wind isn't blowing. Converting renewables to hydrogen and then back to electricity is so costly and inefficient that many other strategies are already more cost-effective, as studies have shown.[8] Also, more efficient long-duration energy storage technologies are now emerging.

The same analysis applies to other key sectors when electricity goes head to head with hydrogen. One European expert reviewed thirty-two independent studies on hydrogen heating from the European Union, the United Kingdom, Brazil, and California. "You need five to six times more electricity to generate the same amount of heat with hydrogen than with an efficient heat pump," he explained. "That's the Achilles' heel for hydrogen. It's basic thermodynamics."[9]

Other highly inefficient uses of renewable electricity, according to a 2023 analysis, are making green hydrogen to directly displace natural gas power generation and "blending hydrogen into the gas grid."[10] Worse, although running hydrogen through a fuel cell produces no pollution, burning it in a retrofitted gas turbine is not clean power. It will generate nitrogen oxides, which form tiny particulates that can penetrate deep into the lungs, leading to serious diseases. And blending safe amounts of hydrogen into the gas grid is an almost complete waste of renewables because it involves a large misallocation of resources while barely reducing GHG emissions.

This "opportunity cost" issue is crucial yet too often absent from media and policy discussions about hydrogen. If we misallocate all those renewables, we lose the opportunity to achieve far more reductions for far less money by using them to directly replace fossil fuels. At the rate the nation and the world are cutting emissions—not fast enough and not at all, respectively—green hydrogen won't make sense until after

2050. Opportunity cost matters, especially because we have only one opportunity to avert catastrophic climate change over the next decade and don't have time, money, and resources to waste.

Second, no one will make a big investment in a hydrogen delivery infrastructure until it's clear that a hydrogen-using technology (like fuel cell cars) is a winning product. But no one will scale up making such hydrogen tech without a vast infrastructure already in place to move hydrogen around cheaply and efficiently. This "chicken and egg" problem of who goes first and risks losing billions has never been solved.

Third, hydrogen is a hazardous fuel to use, and in industrial settings, a great deal of money and training are spent to avoid serious accidents. As a 2021 "Review on Hydrogen Safety Issues" noted, "It is difficult to maintain safety in production, storage, transportation, and utilization because of the inherent characteristics of hydrogen."[11] Hydrogen has many combustion regimes, as a major 2011 report explained.[12] For hydrogen concentrations above 10%, acceleration of the combustion wave up to sound velocity (about 1,000 feet per second) "has been found in many experiments." But scenarios with higher concentrations can create a "detonation," which is "a combustion wave that travels at supersonic speeds relative to the unburnt gas in front of it." That speed can be over 6,000 feet per second.

Fourth, there is still no practical and affordable way to store hydrogen, which makes transporting it over long distances expensive and energy intensive. Many companies want to liquefy and supercool hydrogen to −423°F (−253°C) for transport, even though that process can consume as much as 40% of the hydrogen's energy. Keeping the temperature that low in a tanker truck is challenging, especially as the liquid hydrogen sloshes back and forth, so it starts to "boil off," dangerously increasing the pressure in the tank. The excess hydrogen is typically vented into the air.

That brings us to one of the biggest challenges hydrogen faces: Even though it is perhaps the leakiest gas, we now know that leaking or releasing hydrogen into the air makes the planet warm much faster—especially in the near term—than scientists realized.

Hydrogen as a Climate Solution

Our energy choices are inextricably tied to the fate of our global climate. The burning of fossil fuels emits CO_2 into the atmosphere, where it builds up, blankets the planet, and traps heat, accelerating global warming. Earth's atmosphere now contains more CO_2 than at any time in the past few million years, leading to rising global temperatures, more extreme weather events (including floods and droughts), devastating wildfires, ocean acidification, sea level rise, and many more threats to humanity.

Failure to rapidly reverse that trend will lead to ever-worsening and ultimately catastrophic climate impacts that will be irreversible for generations. In the past two decades, the science has grown stronger that to avoid such impacts, the world needs to hold total warming to "well below 2°C above pre-industrial levels" while "pursuing efforts to limit the temperature increase to 1.5°C," as the nations of the world agreed unanimously in the 2015 Paris Agreement and reaffirmed in Glasgow in 2021.

The world's top climate experts of the Intergovernmental Panel on Climate Change explained in their 2018 *Special Report on Global Warming of 1.5°C* that meeting the 1.5°C (2.7°F) limit requires net zero global emissions by 2050, whereby whatever emissions the world can't mitigate would be offset by negative emission technologies that currently aren't commercial.[13] Keeping warming below 2°C (3.6°F) still requires net zero global emissions just a decade or two later.

Yet since 2000, annual global emissions have risen to the equivalent of 50 billion tons of CO_2. In fact, we are currently headed to as much as 6°F (3.3°C) warming by 2100 based on policies in place globally.[14]

However, our urgent need to rapidly reduce the rate of warming is at odds with the rapid expansion of hydrogen use because the latest science reveals hydrogen is a major indirect GHG. Hydrogen is not a GHG that directly traps heat, as carbon dioxide and methane do, "but its chemical reactions change the abundances of the greenhouse gases methane, ozone, and stratospheric water vapor, as well as aerosols," explains a 2023 journal article.[15]

The result is that hydrogen has thirty to forty times the global warming potential of CO_2 over a twenty-year period (GWP20), a critical period for reaching the goals of the Paris Agreement. Unfortunately, recent studies tend to overhype the climate benefit of hydrogen: Not only do they use the much lower hundred-year GWP of hydrogen (GWP100), but they also use outdated estimates. That approach can make hydrogen strategies that actually increase net CO_2 emissions look as if they offer some climate benefit because the new GWP20 of hydrogen is about seven times bigger than the old GWP100.[16]

The authors of the 2023 article note that "By reducing short-lived climate forcers, such as hydrogen, there is potential to slow down global warming in the next 20–25 years." Unfortunately, we will be doing the exact opposite. The study urges, "It will be important to keep hydrogen leakages at a minimum to accomplish the benefits of switching to a hydrogen economy." Yet hydrogen is the most leak-prone gas because it is the tiniest and lightest of molecules. Many current proposals imagine vastly more hydrogen production, transportation, and use, all but guaranteeing a massive surge in hydrogen leaks and venting of boil-offs from liquid hydrogen storage.

Significantly, hydrogen leaks are very hard to detect, and undetected hydrogen leaks cause 22%–40% of hydrogen accidents.[17] Until and unless we can drive hydrogen leaks and boil-offs down to near zero, it will make no sense to divert renewable energy to its production.

If the world had started cutting emissions when we started having annual meetings to address the problem—three decades ago—things would be much easier. But during the time we have had nearly thirty global climate meetings, GHG emissions have risen 50%. Also, we now know that to avert the worst impacts, we must reduce emissions as close to zero by midcentury as we can, which means we need to have solutions that can get very big, very fast.

Analysis by Massachusetts Institute of Technology researchers working with the nonprofit Climate Interactive has found that technologies such as fusion, which still need at least one breakthrough to become commercial, are unlikely to contribute to lowering global temperatures in 2100 by more than 0.1°C. And it's entirely possible that fusion—the most unique and challenging use of hydrogen-based energy—doesn't even achieve that.

Long-term research and development of strategies such as hydrogen as an energy carrier or direct air capture (DAC) and storage of CO_2 remain worthwhile. But both need major breakthroughs before significant money should be devoted to scaling them up. That's especially true because DAC has the same major flaw as hydrogen: Its overall efficiency is low (5%–10%) but the price is high.

So both hydrogen and DAC consume vast amounts of renewables that could achieve far more CO_2 emission reductions for far less money if used to directly replace fossil fuels. Also, it's entirely possible, if not likely, that there will be better and cheaper ways of doing what they can do. Hyping them as major pre-2050 climate solutions is counterproductive, and subsidizing significant deployment of either means greatly misallocating renewable energy and money—a big opportunity cost.

Because the goal of hydrogen is replacing fossil fuels, "the electricity you use to make it would have to be ridiculously cheap. And if you have that, why use it to make hydrogen?" as inventor and electrification expert Saul Griffith told the *New York Times* in 2023.[18] In 2022, Griffith explained, "I developed hydrogen tanks that I sold to a consortium of the world's auto companies," and so "I will do very well if hydrogen is a winner."[19] It's his expertise with hydrogen that fuels his skepticism.

Our decades of dawdling mean we must focus the overwhelming majority of our money on rapidly deploying technologies that are already successful, such as solar power, wind power, batteries for storage, electric vehicles instead of gasoline ones, and electric heat pumps instead of natural gas boilers. In every sector, we must pour more money into all electrification technologies than we do into hydrogen energy technologies because, again, the hydrogen solution to a climate problem is exceedingly unlikely to compete with the electric one.

International Energy Agency energy analyst Dr. Simon Bennett, who coauthored their 200-page *Future of Hydrogen* report in 2019, told me in 2024, "Many of the technologies that have generated the most excitement in the past decade were ones that had the effect of outcompeting hydrogen or CCS [carbon capture and storage] in an application where we used to think hydrogen was the best approach."

The (New) Hype About Hydrogen

Unfortunately, the fossil fuel industry—which generates hundreds of billions of dollars annually in revenues from promoting the fuels destroying our livable climate—sees an opportunity to make tens of billions from hydrogen while continuing to produce polluting fossil fuels. So it has spent millions promoting the idea. As Canary Media explained in October 2023, fossil fuel companies "form the core membership of the biggest hydrogen lobbying groups."[20] They are one of the few industries that use hydrogen directly—and the only one that produces the fossil fuels responsible for making virtually all hydrogen today.

The industry believes that the challenges of making hydrogen with renewables (or nuclear power) mean it will inevitably be made by reforming natural gas and then capturing and storing the resulting CO_2, so-called blue hydrogen. But as one 2023 analysis concluded, "Blue hydrogen is not clean or low-carbon and never will be."[21] Also, as Bloomberg warned in a 2021 article, it has inherent economic limitations.[22]

The fossil fuel industry isn't the only one overhyping hydrogen and other similarly sexy strategies that are implausible or even counterproductive as near-term solutions. The Department of Energy promotes all energy technologies and seldom criticizes any of them, no matter how dubious or implausible. The US Inflation Reduction Act alone could lead to a half trillion dollars or more in tax credits for hydrogen that either achieve very expensive GHG reductions or actually increase net emissions.

The venture capital community also oversells its investments, pouring billions of dollars into every conceivable technology without distinguishing between those with high and low potential for impact. The companies themselves are on an endless quest to raise money and convince venture capitalists they can deliver on their promises of large, rapid, and affordable emission reductions, no matter how implausible.

Major media outlets, from *The Economist* to the Associated Press to the *New York Times*, write credulously about products that still need at least one breakthrough to be affordable and scalable—and thus are unlikely to be major contributors to the emission reductions we need by 2050. Throughout the book, I'll document some of this hype because it has become so deceptively pervasive.

Much of the bad policy and overhype driving this resurgence of interest in hydrogen stems from a broad lack of understanding about the intractable nature of hydrogen as an energy carrier, the inherent inefficiency of carbon-free hydrogen, and its inability to compete with electricity, especially given rapid advances in electrification technologies. These challenges are the major focus of this book.

Chapter 1 focuses on the history of hyping hydrogen as an energy carrier from the 1990s through today. Chapter 2 looks at the much longer history of hydrogen itself and what might be called the original hype about hydrogen: fusion energy, which is sometimes described as "the future of energy, and it always will be."

In Chapters 3 through 5, I'll dive into hydrogen production. Chapter 3 provides an overview of how hydrogen is produced today and how it might be produced with little or no carbon emissions in the future. It explains why green hydrogen is by far the most important type of hydrogen in our rapidly warming world but is hard to make affordably. It also examines why geologic hydrogen is not likely to be practical, plentiful, or pollution-free.

In Chapter 4, we'll see why nuclear power is far too expensive and impractical to scale up hydrogen production and dig into the safety concerns with hydrogen generation near nuclear reactors. We'll also examine why "economics killed small nuclear power plants in the past—and probably will keep doing so," as the *IEEE Spectrum* presciently warned in 2015.[23]

Chapter 5 examines the hydrogen production method favored by fossil fuel companies: adding CCS to a traditional system for reforming natural gas into hydrogen. It explains why this approach leads to high GHG emissions and why, if we ever develop a practical and affordable CCS system, we should just use it directly on a fossil fuel power plant.

Chapters 6 and 7 explore some of the major applications for hydrogen currently being funded. In Chapter 6, we'll look at the overhyped, complex, expensive, and inefficient technology of direct air capture of CO_2, which many seek to combine with hydrogen to make synthetic liquid hydrocarbon "e-fuels." Although e-fuels are often touted as solutions for air travel and shipping—and even ground transport—we'll look at their many fatal limitations, including their huge opportunity costs.

Chapter 7 details another proposed application—hydrogen cars— and why electric cars have crushed hydrogen ones over the past two decades. We'll see how hydrogen's immutable limitations undercut the business case for it as an energy carrier in virtually all applications.

In the Conclusion, I'll review the book's central arguments and examine how we can accelerate real climate and energy solutions while avoiding the huge opportunity cost of the many false promises.

The Future of Hydrogen

Two decades ago, I ended the Introduction to this book as follows: "This book makes the case that hydrogen vehicles are unlikely to achieve even a 5% market penetration by 2030. And this in turn leads to the book's major conclusion for all readers, from policymakers to corporate executives to investors to anyone who cares about the future of the planet: *Neither government policy nor business investment should be based on the belief that hydrogen cars will have meaningful commercial success in the near- or medium-term.*"

The prediction was accurate. Indeed, it's exceedingly unlikely hydrogen vehicles will achieve even a 1% market penetration by 2030. In fact, 1% by 2050 would be an achievement at this point given the remarkable and ongoing progress of both electric vehicles and batteries.

The original prediction can now be broadened, and perhaps surprisingly, it's not pessimistic—it is merely realistic. Making enough green hydrogen to replace the 100 million tons of hydrogen used directly as a chemical feedstock today (primarily in the fertilizer and petrochemical industries) would be a worthy and meaningful focus of our hydrogen effort. That could save up to 2% of global emissions—if it is truly green. Beyond that, hydrogen energy technologies are unlikely to contribute even 1% of overall net GHG emission reductions in 2050.

That's especially true when we factor in the warming impact of hydrogen leaking and venting. Also, scaling up green hydrogen before most fossil fuel emissions have been eliminated would be counterproductive because of the huge opportunity cost. It would misallocate enormous amounts of renewables that could achieve far greater emission reductions at far lower costs elsewhere in the economy.

For instance, one of the most highly touted applications for hydrogen is a part of the steelmaking process responsible for about 2% of global emissions. But achieving that in the next quarter century would require a massive and expensive industry retooling—"many hundreds of billions of dollars," according to one industry expert.[24] The cost per ton of CO_2 saved would probably be much higher than fully electrifying sectors such as ground transportation and heat for buildings and industry. So while it's worth pursuing the technology's development and demonstration, it shouldn't be a priority for near-term rapid deployment. The best near-term strategy for steel is to focus on increasing the recycling rate of scrap steel, which can be converted to "secondary" steel in an electrified process with an electric arc furnace. At the same time, we should increase funding for development and demonstration of emerging technologies that make primary steel directly with electricity. Beyond steel, most other plausible applications are niche areas with lots of competition, as we'll see.

That reality, in turn, leads to this book's major conclusion for all readers, from policymakers to corporate executives to investors to anyone who cares about the future of the planet: *Neither government policy nor business investment should be based on the belief that hydrogen energy technologies will inevitably be even a small contributor to decarbonizing the energy system by 2050.*

In other words, hydrogen investments do not deserve a privileged position among the list of potential climate solutions. Quite the reverse. Nothing about hydrogen justifies hundreds of billions of dollars in tax incentives or subsidies over the next fifteen years in the United States or worldwide.

Hydrogen is far more likely to be a major boondoggle than anything else, but most people in government, business, venture capital, and the media apparently do not understand that reality. Therefore, it's important to understand how hydrogen works, what its risks and opportunities are, why it has utterly failed to live up to its promise as an energy carrier and climate solution despite considerable effort and money—and the growing number of reasons it remains a bad bet to do so for decades to come.

Finally, while the outcome of the 2024 US Presidential election does not change any of the analysis here, it does strengthen the conclusions. The winner has repeatedly called climate action a hoax, has spread misinformation about core climate solutions like wind power and batteries, and is an avowed opponent of domestic and international climate action. This means it will be much harder for the nation and the world to achieve the Paris climate targets—which another decade of dawdling has already put in grave jeopardy—and avoid catastrophic climate change.

The Trump Administration is likely not only to slow or even terminate domestic policies aimed at reducing GHG emissions but also to actively promote policies that will increase them. So it will become even more important for the rest of us—including states, companies, investors, nonprofits, other countries, and all those who care about maintaining a livable climate—to achieve the highest possible emissions reductions with our actions and money. We must redouble our focus on the real electrification solutions. We simply cannot afford the wasted time and opportunity cost of the many false promises I have discussed here as the window to keeping warming below 3.6°F (2°C) starts to shut even faster.

CHAPTER 1

Hydrogen Hype—Then and Now

From William Grove's first hydrogen fuel cell to today, billions and billions of dollars have been spent attempting to refine and scale hydrogen technology. Throughout this book, we'll see why efforts to make hydrogen a dominant energy carrier have largely failed—and why they're even less likely to succeed now than two decades ago, when I wrote the first edition of this book.

Yet we are now in a frantic rush to spend tens, if not hundreds, of billions of dollars on hydrogen. Trying once again to advance hydrogen as a significant energy solution to the climate problem at the expense of real solutions is probably one of our biggest climate policy blunders, in the United States and worldwide.

To understand what hydrogen's limitations are and how we are just repeating past errors, we need to look back at modern attempts to develop it. My hydrogen journey started in the early 1990s, coinciding with a resurgence of interest in hydrogen as a possible energy carrier or car fuel.

The 1990s: A Hydrogen Fuel Cell Breakthrough?

I first came to the US Department of Energy (DOE) in 1993 to help oversee research and development (R&D) in clean energy. At that time, hydrogen R&D did not even have its own separate budget line but instead was nestled inside the renewable energy budget. For the previous decade, hydrogen research funding had languished in the $1–$2 million per year range, one-hundredth of 1% of the overall departmental budget—a penny out of every $100. Also, annual DOE funding for the fuel cells needed for vehicles—proton-exchange membrane (PEM) fuel cells—was less than $10 million.

Within days of my arrival, I was briefed on recent work by one of the national laboratories that DOE oversaw. They'd had a breakthrough with the potential to dramatically reduce the cost of PEM fuel cells.

Fuel cells are one of the Holy Grails of energy technology. They are pollution-free electric "engines" that run on hydrogen. Unlike virtually all other engines, fuel cells do not rely on combustion or fossil fuels. Therefore, they produce no combustion byproducts, such as oxides of nitrogen, sulfur dioxide, or particulates, the air pollutants that cause smog and acid rain and that have been most clearly documented as harmful to human health.

Fuel cells have reliably provided electricity to spacecraft since the 1960s, including the Gemini and Apollo missions and the space shuttle. But finding a fuel cell with the right combination of features for powering a car or truck proved much more difficult. Why? To begin with, you need a fuel cell that is lightweight and compact enough to fit under the hood of a car but that can still deliver the power and acceleration drivers have come to expect. The fuel cell must also be able to reach full power in a matter of seconds after startup, which rules out a variety of fuel cells that operate at very high temperatures and thus take a long time to warm up. Fuel cells must also be similar in price and reliability to the gasoline-powered internal combustion engine, an exceedingly mature technology that is the product of more than a hundred years of development and real-world testing in hundreds of millions of vehicles.

Furthermore, these hardware hurdles are all quite separate from how the fuel for fuel cells—hydrogen—would be produced and delivered to the vehicle. Hydrogen, first and foremost, is not a primary fuel, like natural gas or coal or wood, which we can drill or dig for or chop down and then use at once. Hydrogen is the most abundant element in the universe, true enough. But on Earth, it is bound up tightly in molecules of water, coal, natural gas, and so on. To unbind it, a great deal of energy must be used. We'll dig much deeper into the implications of these properties of hydrogen later in the book.

For all these reasons—plus the sharp drop in government funding for alternative energy under president Ronald Reagan—hydrogen fuel cell vehicles (FCVs) received little attention through most of the 1980s. Still, a few government and industry laboratories (along with a small and ardent group of hydrogen advocates and state energy experts) kept plugging away, particularly on PEM fuel cells.

Fuel cells require catalysts to speed up the electrochemical reaction, and PEM fuel cells use platinum, a very expensive metal. An early 1981 analysis for the DOE presciently argued that PEM fuel cells would be ideal for transportation if the amount of platinum could be sharply reduced. By the early 1990s, Los Alamos National Laboratory (and others) had succeeded in cutting the amount of platinum by almost a factor of ten, a remarkable improvement. This still did not make PEM fuel cells close to cost-competitive with gasoline engines, but it did dramatically reinvigorate interest in hydrogen-powered vehicles. PEMs were exactly the kind of low-temperature fuel cell that could be used in a car.

After I was briefed on the Los Alamos work at DOE, I began urging increases in funding for PEM fuel cells and for the development of a transportation fuel cell strategy. At the time, it was far from clear that electric vehicles (EVs) could satisfy all the criteria for a successful alternative fuel vehicle (AFV) to replace liquid fossil fuels. Hydrogen FCVs were the only plausible AFV that could be CO_2-free and scalable.

The Clinton administration's entire team supported R&D for fuel-efficient technologies. So funding for hybrid vehicles, including fuel cells, increased significantly. So, too, did funding for hydrogen production, because even if we succeeded in commercializing fuel cells, we would still need to find affordable, pollution-free sources for hydrogen. Even today, over 95% of hydrogen is made from fossil fuels—coal, oil, and especially methane (CH_4), which releases vast quantities of CO_2, the primary human-caused greenhouse gas.

In mid-1995, I moved to the DOE's billion-dollar Office of Energy Efficiency and Renewable Energy, where I was in charge of all budget and technology analysis. Ultimately, I ran the entire office for much of 1997 as acting assistant secretary. I was able to work with other fuel cell advocates in and out of the administration, especially my DOE colleagues Brian Castelli and Christine Ervin, to keep the PEM fuel cell budget creeping upward even as the entire budget for the office was cut 20% by a 1995 Congress that was extremely skeptical of all energy R&D.

By 1998, when I left DOE, the hydrogen budget was ten times larger than in 1993, and the proposed PEM fuel cell budget was more than three times larger. This funding did help drive down the cost of fuel cells (PEM and others) steadily while performance increased. Harry Pearce, vice chairman of the General Motors Corporation, said at the North American International Auto Show in January 2000, "It was the Department of Energy that took fuel cells from the aerospace industry to the automotive industry, and they should receive a lot of credit for bringing it to us."[1]

The 2000s: Hyping Hydrogen

In the 2000s, hydrogen budgets were ballooning everywhere in a race for patents and products. Hydrogen hype was also ballooning. William Clay Ford Jr., chairman of the Ford Motor Company, said in October 2000, "I believe fuel cells will finally end the 100-year reign of the internal combustion engine," a poignant statement from the great-grandson of the man whose manufacturing innovations had begun that reign a century earlier with the Model T. But he was wrong.

This was the era of President Bush's 2003 State of the Union, in which he called for a historic $1.2 billion in research "so that America can lead the world in developing clean, hydrogen-powered automobiles." Yet as I began researching and writing the first edition of *The Hype About Hydrogen*, it became clear that more hype and more funding wouldn't be enough to overcome hydrogen's many obstacles.

Toyota had just come out with its game-changing second-generation Prius. It was far more fuel efficient than traditional internal combustion engine cars, and its tailpipe had so little pollution that California classified it as a partial zero-emission vehicle. In the first edition of this book, published in 2003, I noted that "hybrid gasoline–electric vehicles *today*, such as the Toyota Prius, are already much more efficient than traditional internal combustion vehicles, and nearly as efficient as *projected* fuel cell vehicles (assuming fuel cells achieved the performance targets)." So, I concluded, "it is possible, we may never see a durable, affordable, fuel cell vehicle, with an efficiency, range, and annual fuel bill that matches the best hybrid vehicle." That is still true.

But even if they ever could beat the best hybrids, hydrogen cars would still have one unfixable problem: the competition from EVs. It was clear two decades ago that if we could build a practical EV that could be charged directly with carbon-free electricity, it would travel four times farther on the same amount of carbon-free electricity than a hydrogen fuel cell car and thus would not only conserve the use of such clean power but also make the electric car much cheaper to run.

So I predicted that "hydrogen vehicles are unlikely to make a significant dent in US greenhouse gas emissions in the first half of this century." Even in the 2000s, it was increasingly clear that hydrogen would make sense as a climate solution only if that sector could not be electrified. We'll look more closely at the competition between electric and fuel cell cars in Chapter 7.

Around when the first edition of this book was published, a National Academy of Sciences panel released a major report with a variety of important technical conclusions suggesting the challenges ahead for hydrogen cars. For instance, the panel said, "The Department of Energy should halt efforts on high-pressure tanks and cryogenic liquid storage. . . . They have little promise of long-term practicality for light-duty vehicles."[2]

The next month, March 2004, a study by the American Physical Society concluded that "a new material must be discovered"[3] to solve the storage problem. No such material has been found, and the two types of onboard hydrogen storage that the academy rejected are still today the only ones being seriously considered.

Bill Reinert, US manager of Toyota's advanced technologies group, was asked in January 2005 when fuel cells cars would replace gasoline-powered cars. He replied, "If I told you 'never,' would you be upset?"[4]

The 2010s: Return to Reality

By the 2010s, the hype about hydrogen seemed to have given way to reality. Even by September 2008, *The Economist* had written that "the promise of hydrogen-powered personal transport seems as elusive as ever," in an article titled, "The Car of the Perpetual Future."[5] The article continued, "The non-emergence of hydrogen cars over the past decade is particularly notable since hydrogen power has been a darling of governments worldwide, which have spent billions of dollars in subsidies and incentives to make hydrogen cars a reality."

In 2013, the independent research and advisory firm Lux Research released "The Great Compression: The Future of the Hydrogen Economy." Their study concluded that despite billions in R&D spent in the previous decade, "the dream of a hydrogen economy envisioned for decades by politicians, economists, and environmentalists is *no nearer*."[6]

Ford Motor Company aggressively pursued FCVs in the first decade of the twenty-first century. By 2009, their thirty Ford Focus FCVs had, combined, over a million miles on them. Yet after all this work, Ford's *Sustainability Report 2013/14* explained, "extensive DOE analysis has not yet revealed an automotive fuel cell technology that meets the DOE's targets for real-world commercialization, or that maintains proper performance throughout the targeted lifetime while staying within the targeted cost."[7]

The onboard storage problem was never solved. Ford concluded in its 2013/2014 report that because of the "still significant challenges related to the cost and availability of hydrogen fuel and onboard hydrogen storage technology," we need "further scientific breakthroughs and continued engineering refinements" to "make fuel cell vehicle technology commercially viable."

Other companies faced similar problems but decided to start marketing vehicles anyway. Toyota's Mirai FCV started selling in Japan in December 2014—and in the US in August 2015—for about $57,000 before incentives. But Pat Cox, former European Parliament president, said in a November 2014 talk on hydrogen cars that Toyota was probably losing between $66,000 and $133,000 on each car.[8]

During this lost decade for FCVs, their major competitor, plug-in electrics, surged into the marketplace thanks to a once-obscure 2006 startup called Tesla Motors. By 2014, Tesla had become California's largest auto industry employer, according to the US Department of Energy.[9] Its market capitalization (the total value of all its stock) then was more than half that of General Motors.

Indeed, battery technology had improved so much during that decade that Tesla was selling a fully electric vehicle with no gas engine. By the end of 2010, Chevy had released the Volt plug-in hybrid in the United States. At the same time, Nissan released the LEAF, which by the end of 2016 had become the world's top-selling all-electric car, with total sales of over 250,000. Tesla's Model S had sales of more than 150,000.

The 2020s: The Hype Strikes Back

In the past ten years, there have been no major breakthroughs in hydrogen technology that would address the key problems of inefficient and expensive production, inadequate storage technologies, and the lack of infrastructure for transporting and fueling hydrogen.

"The math still doesn't add up," the head of European Union lending and advisory operations at the European Investment Bank told Bloomberg in May 2024.[10] "At the moment, there is no case that we can see for a model based on green hydrogen independently produced and distributed as an energy commodity."

In its *Sustainability Report 2013/14*, Ford devoted two full pages to "Hydrogen Fuel Cell Vehicles (FCVs)," used the term "FCV" thirty times, and featured the Ford Focus FCV. By the time of its *2024 Integrated Sustainability and Financial Report*,[11] Ford devoted only two paragraphs to the subject and never used the term "FCV," and the Ford Focus FCV was replaced by the "fully electric Focus electric vehicle (EV)." Yet Ford hasn't completely given up on FCVs. Today, they have moved on to a partnership with the DOE aimed at developing and demonstrating medium- and heavy-duty fuel cell trucks. But even if such trucks ever became commercial, they are unlikely to beat the electric versions that will be available by then (see the Conclusion).

Indeed, today many manufacturers are looking at using liquid hydrogen for fueling heavy-duty trucks because it's denser than compressed hydrogen.[12] But although that may be more practical in the narrow sense that the truck can carry more fuel and have a longer range, it's unlikely to be good for the climate. As the National Academy of Sciences noted back in 2004, "The process of liquefying molecular hydrogen consumes up to 40% of the energy content of the weight shipped." Cooling hydrogen to –423°F (–253°C), just a few degrees above absolute zero—and keeping that cool both in transport and in the vehicle—is energy intensive. And hydrogen is typically vented into the air to avoid a pressure buildup from evaporation. But now that we know hydrogen emissions warm the Earth far more than we ever realized—thirty to forty times as much as CO_2 does over a twenty-year period—liquid hydrogen makes no sense unless such venting is eliminated.

Meanwhile, the primary competitors to hydrogen cars, EVs, have been rapidly and deeply penetrating car markets worldwide. As predicted, EVs have won the race for the car of the future. Electric cars are outselling hydrogen FCVs by nearly 1,000 to 1. In 2024, EVs hit 20% of new car sales worldwide and 50% in China. The battle for the car of the future is over. It was over before it started.

But this type of one-sidedness is what we should expect in every sector where electricity competes directly with hydrogen. For instance, European countries have been considering hydrogen for heating homes and other buildings even though electric heat pumps are extremely efficient. Heat pumps don't use electricity to generate heat but rather extract heat from outside air, even cold air. So they can provide three to four units of heat for each unit of electricity consumed. Green hydrogen for heating can't compete with heat pumps run on renewables. As one 2023 analyst calculated, "You need five to six times more electricity to generate the same amount of heat with hydrogen than with an efficient heat pump."[13]

And yet despite all this, the hype is back like an unkillable vampire. Even *The Economist* ran a 2020 piece titled "After Many False Starts, Hydrogen Power Might Now Bear Fruit."[14] The article began by talking up the South Korean hydrogen car industry. But in 2023, the country's registration rate of hydrogen cars fell 54% to about 4,600, while more than thirty times as many battery electric cars were sold that year.[15]

Governments have put in place staggering incentives to prop up hydrogen. "Nine of the world's 10 biggest carbon polluters have published hydrogen strategies and incentives to grow the fuel's use, which globally already exceed $360 billion, according to BloombergNEF," reported a 2024 Bloomberg series on "Hydrogen Hype."[16]

Yet that $360 billion figure appears to be a low estimate. Indeed, perhaps the single clearest example of just how wildly overhyped hydrogen is today is the staggering misallocation of money and resources from Section 45V of the 2022 Inflation Reduction Act, which created a tax credit for low-carbon hydrogen. A detailed November 2023 analysis by the Electric Power Research Institute on the impact of this tax credit found that "cumulative fiscal costs through 2040 are $504–756 billion."[17] Yet these funds result in only "modest emissions reductions." The result is "very high fiscal outlays per tonne of CO_2 reduced . . . which can exceed $750/t-$CO_2$. These costs are approximately an order-of-magnitude higher than the implied abatement costs of other [Inflation Reduction Act] credits." That is, the clean hydrogen tax credits are about ten times more expensive per ton of CO_2 reduced than other clean energy tax credits.

But even these minimal results assume the enactment and enforcement of rigorous rules for obtaining the credit (see Chapter 3). If the rules are weakened—a major goal of the oil and gas industry and others—then, as the Electric Power Research Institute wrote, "similar to most power-sector-focused studies, we find that economy-wide CO_2 emissions generally increase with IRA's hydrogen tax credits."

In short, the United States could easily spend half a trillion dollars or more on hydrogen tax credits that have no emission impact and may even make the climate problem worse. Indeed, given hydrogen's warming potential, the inevitability of leaks and boil-off, and the challenges facing green hydrogen, making the climate problem worse seems all but inevitable. The opportunity cost of such a misallocation of resources would be staggering.

No wonder many leading experts have strongly criticized these tax credits, with one describing them in 2024 as somewhere between "an almost complete distraction" and "an almost complete boondoggle."[18] It appears we have learned nothing from the overhyping of hydrogen two decades ago, even though hydrogen's prospects as an energy carrier are even lower than they were back then. I hope this new edition of my book can help finally put an end to this cycle.

CHAPTER 2

Hydrogen History and the Hype About Fusion

Hᴏᴅʀᴏɢᴇɴ ɪꜱ ᴛʜᴇ ʙᴇꜱᴛ ᴏꜰ ꜰᴜᴇʟꜱ. Hydrogen is the worst of fuels.

On the plus side, hydrogen is the most abundant element in the universe by far. Hydrogen can power highly efficient fuel cells that generate both electricity and heat with no emissions other than pure, drinkable water. Hydrogen is one of the few substitutes for fossil fuels that will not contribute to global warming—if generated by carbon-free sources, such as solar and wind power or nuclear energy.

Hydrogen is widely used in industry today, and we have decades of experience generating, storing, and transporting it. It is a very hazardous fuel, but it has a good safety record in industrial settings with strict rules, protocols, and worker training, which can add significant cost and complexity to a major hydrogen project.

On the minus side, hydrogen is not an energy source; it is an energy carrier, like electricity. It is not readily accessible in vast, highly accessible quantities, as are coal, oil, natural gas, sunlight, wind, and even electricity. Many companies are pursuing so-called natural or geologic hydrogen, which is found naturally beneath the earth's surface, but at this point, it is highly unlikely that will prove a game changer (see Chapter 3).

Hydrogen can theoretically power a reactor through fusion, the way the sun does. But after decades of research and hype, fusion is unlikely to contribute to lowering global temperatures in 2100 by more than 0.1°C. Indeed, many experts believe we may still be decades away from achieving all the breakthroughs needed for fusion to be a winning and scalable climate solution. The challenges fusion faces in becoming a climate solution are worth understanding because the overhyped solutions examined in this book face many of the exact same challenges.

Hydrogen is bound up tightly in such molecules as water and natural gas, so it is expensive and energy-intensive to extract and purify. Fuel cells have proven remarkably resistant to profitable commercialization after many decades of trying. The hydrogen most needed for fighting climate change is zero-carbon "green hydrogen" from renewable sources. But it is especially expensive, and the primary reason it has gotten cheaper is that renewable electricity has gotten much cheaper. Yet that is little help to the mass commercialization of zero-carbon hydrogen because technologies that run directly on renewable electricity are the main competitor to technologies that might run on hydrogen made from renewable electricity.

Similarly, one of the most overhyped strategies is "blue hydrogen"—hydrogen from natural gas where the resulting CO_2 emissions are captured and stored. Neither the practicality nor the scalability of blue hydrogen has been proven, whereas several studies show that it would not be close to carbon free and could increase global warming.

Hydrogen is difficult and costly to compress, store, and transport. It has a very low energy per unit volume, one third that of natural gas. It is expensive and energy intensive to liquefy; moreover, a gallon of liquid hydrogen has only about one quarter the energy of a gallon of gasoline. Hydrogen has major safety problems: It is flammable over a wide range of concentrations and has an ignition energy one twentieth that of natural gas or gasoline, so leaks are a significant fire hazard. Because it is the most leak-prone gas, hydrogen is subject to a set of strict and cumbersome codes and standards. And we now know hydrogen has a major heat-warming ability as an indirect greenhouse gas. Any significant leaks would undermine if not eliminate entirely the modest value of even green hydrogen as a climate solution.

To paraphrase Charles Dickens again, we were all going directly to hydrogen, we were all going directly the other way. This chapter looks at options for producing hydrogen today and in the future. But first a little background on this unique element.

A Brief History of Hydrogen

"If God did create the world with a word, the word would have been hydrogen," said Harlow Shapley, one of the twentieth century's greatest astronomers. In our current theory of the origin of the universe, when atoms first formed out of the sea of particles created in the Big Bang, basic hydrogen—one electron and one proton—constituted some 92% of the atoms, and virtually all the rest was helium. Today, some 15 billion years later, about 90% of all particles are hydrogen and 9% are helium.

Hydrogen is everywhere. In the near-vacuum of interstellar space, every cubic centimeter contains a few hydrogen atoms. In the interior of planet Jupiter, every cubic centimeter contains more than 10 million billion billion hydrogen atoms. Every second, our sun fuses 600 million tons of hydrogen into helium, providing the light and warmth that make life on Earth possible. The hydrogen-driven process of stellar birth and death has generated all the heavier elements that make up our bodies and our planet, including the oxygen that, with hydrogen, composes water. As physicist John Rigden puts it in his delightful 2002 book *Hydrogen: The Essential Element*, "hydrogen is the mother of all atoms and molecules."[1]

Hydrogen's first "discovery" on Earth is often credited to Theophrastus Bombastus von Hohenheim (1493–1541), better known as Paracelsus, Renaissance physician and alchemist. Paracelsus observed that when acids react with metals, a flammable gas is emitted. But English nobleman Henry Cavendish (1731–1810) is typically given credit for discovering in 1766 that hydrogen is a separate substance and for characterizing a number of its qualities. He called hydrogen "inflammable air" and was the first chemist to produce water from oxygen and hydrogen. French scientist Antoine Lavoisier (1743–1794) learned of Cavendish's work in 1783 and repeated and expanded upon his experiments. Lavoisier, who died on the guillotine in 1794, named the element *hydrogène*, from the Greek words for "water" and "generate."

How remarkable, then, that hydrogen not only generates water when combined with oxygen but also helped generate the oxygen in stars in the first place. How doubly remarkable it would be if the sun's hydrogen-generated power were ultimately used to electrolyze that water to generate hydrogen to power the planet.

The idea of hydrogen as the ultimate fuel, limitless and powerful, stoked the imagination of nineteenth-century fiction writers. In his 1874 book *The Mysterious Island,* the ever-prescient writer Jules Verne has his characters discuss what would happen to America's "industrial and commercial movement" when the world runs out of coal in hundreds of years. The engineer, Cyrus Harding, explains that the world will turn to another fuel, "water decomposed into its primitive elements . . . doubtless by electricity, which will then have become a powerful and manageable force." Harding goes on to say,

> Yes, my friends, I believe that water will one day be employed as fuel, that hydrogen and oxygen which constitute it, used singly or together, will furnish an inexhaustible source of heat and light, of an intensity of which coal is not capable. Some day the coalrooms of steamers and

the tenders of locomotives will, instead of coal, be stored with these two condensed gases, which will burn in the furnaces with enormous calorific power. There is, therefore, nothing to fear. As long as the earth is inhabited it will supply the wants of its inhabitants, and there will be no want of either light or heat as long as the productions of the vegetable, mineral or animal kingdoms do not fail us. I believe, then, that when the deposits of coal are exhausted we shall heat and warm ourselves with water. *Water will be the coal of the future.*[2]

Although Verne neglected to tell us where the primary energy to electrolyze the hydrogen would come from, his prediction is remarkable.

A few years later, in 1893, British novelist Max Pemberton published a prophetic bestseller, *The Iron Pirate*. In the book, a speedy and powerful pirate ship terrorizes the Atlantic Ocean. The source of its power baffles the narrator: "Neither steam nor smoke came from her, no evidence, even the most trifling, of that terrible power which was then driving her through the seas at such a fearful speed." Ultimately, we learn that the ship is based in Greenland, and hydrogen from coal fuels "the most powerful engines that have yet been placed in a battleship." Interestingly, Pemberton writes, "the gas itself was made by passing the steam from a comparatively small boiler through a coke and anthracite furnace, the coke combining with the oxygen and leaving pure hydrogen."[3] Coal is still a major source of hydrogen, and it remains a significant area of focus for public and private sector research and deployment.

As the twentieth century dawned, hydrogen became an intense focus of scientists, both theorists and experimentalists. "To understand hydrogen is to understand all of physics," one physicist said. The study of hydrogen was critical to our understanding of the atom and the universe's evolution, to the development of quantum mechanics and quantum electrodynamics, and to such practical devices as magnetic resonance imaging medical equipment.

Hype and Confusion About Fusion

In the mid-twentieth century, scientists and governments spent tremendous effort trying to replicate on Earth the awesome power of the sun's hydrogen fusion. Strikingly, we have been all too successful in tapping fusion energy in uncontrolled reactions—unleashing the horrific power of the hydrogen bomb—but unsuccessful in efforts to create practical energy sources from a controlled fusion reaction to generate electricity.

Few science and engineering challenges have proven as intractable as trying to sustain the temperature of 100,000,000°C (180,000,000°F) needed to fuse hydrogen atoms together, capture the resulting energy released, generate electricity, and keep the reaction running as reliably, safely, and cheaply as a typical power plant. That's why fusion is still far from joining the growing ranks of useful climate solutions.

Since the 1950s, the US government has spent over $30 billion on fusion energy science. Media hype was common. A January 1982 *New York Times Magazine* article about scientists working on fusion asserted, "Now, more than 30 years after they began, it appears that these fusion scientists will, indeed, be successful."[4] Yet just a year later, when I was a graduate student in physics at the Massachusetts Institute of Technology, the MIT-edited magazine *Technology Review* published a bombshell cover story by MIT professor Lawrence Lidsky, "The Trouble With Fusion."[5]

The 1983 cover had a black background and this sentence from the article in quotation marks: "Even if the fusion program produces a reactor, no one will want it." The problem was that the reactor fuel was deuterium and tritium, hydrogen isotopes with one and two extra neutrons, respectively. "The most serious difficulty concerns the high-energy neutrons released in the deuterium–tritium reaction," the article explained. "These particles damage the reactor structure and make it radioactive.

A chain of undesirable effects ensures that any reactor employing D–T fusion will be a large, expensive, and unreliable source of power."

What made this critique doubly devastating was that Lidsky was then associate director of the Plasma Fusion Center and editor of the *Journal of Fusion Energy*. Lidsky was writing about "the most likely scheme," where a tubular reactor is curved to form a doughnut while powerful magnetic fields are used to control the reaction. This approach is known as magnetic confinement.

It took four decades for excitement in fusion to be reignited via a very different but less practical approach: inertial confinement. In December 2022, the National Ignition Facility (NIF), a research device that uses the largest and highest-energy laser in the world to confine the high-temperature reaction, very briefly sustained a fusion reaction at Lawrence Livermore National Laboratory near San Francisco. But what happened at Livermore was wildly overhyped, which is worth examining since hype has become the norm these days for technologies that need one or more breakthroughs to even have a shot at being practical solutions.

The actual efficiency of the system—total electricity generated from the fusion divided by total electricity input—was about 1%. That is, the NIF consumed 100 times as much power from the grid as it produced.

If you thought that there was actually net energy gain—more electricity coming out than put in—that was the result of the confusing and misleading way in which the results were presented.

Construction was started on the NIF back in 1997, when I was still at the Department of Energy (DOE). The NIF had been hotly debated in the DOE for a few years since the department and its predecessors had spent staggering sums on fusion with little progress. At the time—and still today—laser fusion was not seen as the likely winner in the race to be a commercial power generation source. After all, the "NIF's 192 powerful laser beams, housed in a 10-story building the size of 3 football fields" deliver their energy in "ultraviolet laser energy in billionth-of-a-second pulses onto a target about the size of a pencil eraser."[6] The system was far from being practical for many reasons other than size.

The NIF almost certainly would not have been approved if its only purpose was to try to make fusion a viable energy source. It was justified largely as a device that could help simulate the explosion of a thermonuclear bomb since the actual detonation of such bombs had been banned decades earlier. Advocates inside and outside DOE argued that without nuclear testing, we couldn't be certain of the reliability of our weapon stock.

"NIF scientists conduct experiments necessary to ensure America's nuclear weapons stockpile remains safe, secure, and reliable without underground testing," explains Livermore's FAQ. NIF is the only DOE facility "with the potential to duplicate all the phenomena that occur in the heart of a modern nuclear device."

Many people at DOE didn't see that as an especially compelling argument, but national security arguments tend to trump all others in government. So the NIF was approved.

Twenty-five years after it was started, the NIF did achieve a genuine breakthrough, although it was much more of a scientific breakthrough than an energy breakthrough. The energy in the laser pulse that ignited the fusion reaction in December 2022 was slightly less than that in the resulting fusion reaction. That meant there was a slight "gain" in the actual reaction that took place. But the lasers are not very efficient in converting electricity into laser blasts, so it took 100 times as much electricity from the electric grid going into the lasers to get that pulse out in the first place. The major media coverage almost entirely missed this important point, but the trade press did better.

Indeed, the *IEEE Spectrum*, the leading publication of the Institute of Electrical and Electronics Engineers, even noted in its December 2022 article, "Fusion 'Breakthrough' Won't Lead to Practical Fusion Energy," that the "NIF has previously pulled some energy accounting sleight of hand."[7]

The article notes that this event was "just one tiny step on the long and unclear road to commercial fusion power. . . . In particular, there are five major extrapolations—and they are *major*—highlighted by the experts IEEE Spectrum spoke with" (emphasis in original). For instance, the NIF fires the laser only several times *a year* but ultimately may need to fire many times a second. Yet the lasers are so powerful that "they damage their own guiding optics every time they fire." The precision required is astonishing: "The research team would sweat laser-alignment discrepancies on the order of trillionths of a meter and time discrepancies of 25 trillionths of a second."

At the same time, costs for the system must drop sharply. Each NIF hydrogen capsule currently costs hundreds of thousands of dollars and can take many months to produce. Precision is vital. Performance can be affected by "imperfections the size of a single bacterium."

If all those problems are solved, we still have the efficiency problem. The reaction must be 100 times or more as effective to make up for the inefficiency of the laser, as noted above. If that problem gets solved, the energy produced from the fusion must still be turned into usable electricity. Proposed designs use a "blanket" to absorb ejected neutrons to generate heat and power a steam turbine. "We've made mock-ups, but we've never made a working blanket," explains Steven Cowley, director of the Princeton Plasma Physics Laboratory.

"Lastly and most crucially," if all those concerns are solved, "it's still not obvious whether inertial fusion can be commercially competitive." This is a core concern. Today, we aren't even close to commercial success here.

Any technology that requires one or more technological break-throughs to be practical, affordable, and scalable has extra hurdles to jump to be a commercial success. Let's understand how this applies to fusion energy because the same concern also applies to hydrogen technologies, small nuclear reactors, direct air capture, and other would-be climate solutions.

Can Fusion Ever Be a Commercial Success?

We've spent a great deal of money trying to achieve the breakthroughs needed to turn fusion into a practical, competitive, and scalable source of carbon-free energy on Earth. In 2006, for instance, China, the European Union, India, Japan, Russia, South Korea, and the United States agreed to fund and run the International Thermonuclear Experimental Reactor, which uses magnetic confinement. It was originally expected to cost some $5–$6 billion to construct and achieve full fusion in the early 2020s.[8] Current cost projections are up to four times that—$20 billion—and the new target date is 2035.

In the last few years, the venture capital community has begun pouring billions of dollars into fusion, even though many of the world's leading independent experts still see little likelihood of success for decades. At the December 2022 press conference for the NIF announcement, Kim Budil, director of Lawrence Livermore National Laboratory, was asked how the experiment changed the common statement that fusion is always decades away. She replied, "Not six decades, I don't think; not five decades, which is what we used to say," but "with concerted effort and investment, *a few decades of research on the underlying technologies could put us in the position to build a power plant.*"

In May 2024, at the Starmus Science Festival, International Thermonuclear Experimental Reactor director-general Pietro Barabaschi gave a talk titled "Can Nuclear Fusion Help to Fuel the World?" The fifty-eight-year-old scientist's answer: "Not in my lifetime."[9]

In fact, there's no guarantee that the breakthroughs needed just to make fusion a practical and sustainable energy source will happen any time soon. For instance, even if we can create a viable fusion reactor that generates more power than we put into it, there is no reason to believe it would be a competitive power source for decades, if ever.

Indeed, even if it ever becomes a competitive power source, its value as an emission reducer would remain small for a long time—unless it were substantially superior to the zero-carbon competition available at the time. That's because if it were only a little better than the competing technologies, all it would be doing is replacing that zero-carbon competition, thereby saving a little money but achieving little or no net emission reductions.

No wonder analysis by MIT researchers working with the nonprofit Climate Interactive found that technologies such as fusion that still need at least one breakthrough to become commercial are unlikely to contribute to lowering global temperatures in 2100 by more than 0.1°C, even if the breakthrough occurs in 2025.[10]

How do we know that it is far from inevitable that nuclear fusion will ever come close to beating the future competition? Consider that the first self-sustaining nuclear chain reaction—the world's first artificial nuclear reactor—was built way back in 1942 by Enrico Fermi. And although nuclear power was once imagined as potentially being "too cheap to meter," now, eight decades later, it is too expensive to compete in most market economies. Just because you can build something doesn't mean it's a winner in the marketplace. France and the United Kingdom found that out when they chose to pursue supersonic air travel, while the United States didn't. Yes, they built the Concorde jet, which operated for a long time. But it never made economic sense, and ultimately the program was terminated.

One of the biggest reasons that new energy technologies achieve success in the market much more slowly than expected is that they underestimate or simply ignore the possibility that the competition is going to get substantially better during the time it takes to bring their new technology to market. That is a major reason renewable energy technologies took so long to achieve their full potential. And as we'll see in Chapter 7, it is one of the many reasons why hydrogen cars were doomed to fail.

The journey to market is a long and perilous one for an energy technology, and there are many opportunities to go awry. In the case of fusion, someone first needs to build a laboratory-based test reactor that can sustain an output of electricity generation in a controlled fashion for an extended period of time that is significantly greater than the electricity put in. Then they will need to build the first large-scale demonstration plant that can run smoothly for far longer. Next, they will need to build the first commercial fusion reactor—something that can be readily manufactured, get licensed by the regulators, be sold in the marketplace, and deliver power reliably. And like most first-of-a-kind plants, it will be very expensive.

If, very optimistically, that occurred by the late 2030s, there will be a lot of competition for providing power when the sun isn't shining and the wind isn't blowing. These include long-term storage of electricity and enhanced geothermal energy (see the Conclusion).

With these technological challenges and growing marketplace competition, the bottom line is that controlled, earthbound fusion is unlikely to matter to the crucial challenge of getting global greenhouse gas emissions as close as possible to zero by midcentury. In the effort to avoid catastrophic climate change, the fusion energy we will probably be tapping will be from the sun, in the form of its renewable radiating energy.

CHAPTER 3

Hydrogen Production: Green and Not-So-Green

THE DREAM OF A HYDROGEN ECONOMY rests first and foremost on finding a practical, affordable, and scalable source of pollution-free hydrogen. After decades of trying, this foundational problem has still not been solved.

In a very real sense, the solution is now farther away than ever. That's because we now know that global warming will speed up if we can't prevent significant leaks in the hydrogen production system—leaks of the hydrogen itself or of the natural gas that is mistakenly touted as part of the solution.

In the popular discussion of hydrogen production, there are many, many other "colors" besides green. One December 2023 article listed a total of fifteen that are in use![1] So keeping track of them all is exceedingly difficult, even for professionals. They do not simplify the discussion because you must always explain what the colors mean each time you use them. In that sense, they seem primarily to serve the purpose of creating the impression that many useful ways to produce hydrogen exist. But they don't. Each option for producing hydrogen is deeply flawed, from the inefficient and expensive to the dangerous and counterproductive.

Green hydrogen from renewables is the only kind that makes sense as a climate solution because the world needs to get close to zero CO_2 emissions by midcentury. But even green hydrogen is useful only if it is made from new renewable power plants in the very specific way discussed below—and if we ignore the opportunity cost of inefficiently using renewables that could have achieved far more emissions reductions for far less money just by directly replacing fossil fuels.

Hydrogen from nuclear reactors is implausible in a world where the cost of nuclear power has been rising for decades in the United States and Europe. Also, integrating a nuclear plant with hydrogen generation only multiplies safety risks. Yet the hype about hydrogen from nuclear plants has risen even faster than their price, as the industry, governments, and venture capitalists have been touting implausible promises about small reactors that defy their long history of failure and cost overruns. So Chapter 4 focuses on hydrogen from nuclear power, whose color is euphemistically called "pink," but mystifyingly, the 2023 article notes that it "can also be called purple, red, or violet."

The only color worth remembering other than green is "blue" hydrogen made from natural gas with carbon capture and storage (CCS). But, as I'll discuss in Chapter 5, it is an irredeemably bad idea that should be called dirty blue hydrogen because it releases substantial amounts of greenhouse gases (GHGs), negating whatever modest benefit hydrogen might have.

The other way of producing hydrogen that has generated much excitement and funding is the possibility of finding large quantities of natural or geologic hydrogen underground. But contrary to the hype that it is the most environmentally friendly type of hydrogen, natural hydrogen is likely to have higher carbon emissions per kilogram than green hydrogen, possibly much higher. Also, as discussed below, it may simply be a dead end.

It's important to remember that finding a scalable, low-cost way to make carbon-free hydrogen won't create a hydrogen economy because it won't solve hydrogen's inherent limitations, such as the difficulty of storing and transporting it without leaks, the lack of existing hydrogen infrastructure, and safety concerns. Any one of these could easily block efforts to turn hydrogen into a practical, affordable, sustainable, and scalable energy carrier. Together, they make hydrogen the most flawed of all the potential climate solutions—yet one on which we're betting tens if not hundreds of billions of dollars.

But if we can't solve the problem of producing and delivering large amounts of affordable carbon-free electricity, the other problems don't matter.

Hydrogen Production: Today and Tomorrow

Hydrogen production is a large, modern industry with commercial roots reaching back more than a hundred years. The United States produces about 10 million metric tons (MT) of hydrogen a year, primarily for two purposes.[2] Petroleum refineries are the biggest users, consuming about 5.5 MT a year in 2021. Hydroformulation, or high-pressure hydrotreating, of petroleum converts heavy crude oils into usable transportation fuels or produces cleaner reformulated gasoline.

The other big use (3.5 MT) is to synthesize ammonia (NH_3) for fertilizer production and methanol (CH_3OH), which can be used as a fuel or a building block for plastics and other products.

Global annual hydrogen production is about 100 million metric tons (100 billion kilograms). About 33% is used for oil refining, 27% for ammonia production, and 27% for methanol.[3]

Hydrogen is produced through a variety of processes using traditional fuels. "Today 95% of the hydrogen produced in the United States is made by natural gas reforming in large central plants," the US Department of Energy (DOE) explains.[4] But only 47% of global hydrogen production is from natural gas, whereas 27% is from coal and 22% from oil (as a refinery byproduct). Only a few percent comes from splitting water with electrolysis, and very little of that is done with new renewable power.

In this chapter and the next two, we'll focus on the most important current types of hydrogen production and possible future alternatives, including hydrogen from natural gas, renewables, nuclear power, and CCS.

Hydrogen from Natural Gas

Natural gas, which is mostly methane (CH_4), is by far the most common source of hydrogen today. Conventional steam methane reforming is a multistep process. In the first step, natural gas reacts with water vapor in tubes that are under high pressure (15–25 atmospheres) and high temperature, 1,300°F to 1,800°F, and that are filled with a catalyst (typically nickel), forming carbon monoxide (CO) and hydrogen:

$$CH_4 + H_2O => CO + 3H_2$$

In the second reaction, called the water–gas shift, the CO is shifted with steam to produce CO_2 and extra hydrogen:

$$CO + H_2O => CO_2 + H_2$$

This reaction can be accomplished in one or two stages at temperature levels ranging from 360°F to 850°F.

The flue gas that leaves the shift reactor is some 70%–80% hydrogen, together with CO_2, CH_4, water vapor, and CO. The hydrogen is then separated from the mixed gas stream for final use, typically via pressure swing adsorption (PSA), achieving high purity. For proton exchange membrane (PEM) fuel cells, a CO removal system is needed, because PEMs tolerate only a few parts per million of CO. Today, the CO_2 is almost always vented to the atmosphere. Both methane reforming and PSA are mature commercial processes. For reforming, the overall energy efficiency (the ratio of the energy in the outputted hydrogen output to that in the inputted fuel) is about 70%.

The cost of producing and delivering hydrogen from reformed natural gas is about $1 to $3 per kilogram, according to BloombergNEF.[5] It depends greatly on the cost of natural gas. The price is closer to the low end of the range in the United States, where natural gas is relatively cheap. In places like Europe, it's much higher.

Of course, in a world fighting to avoid catastrophic climate change, making hydrogen while releasing huge amounts of carbon dioxide into the air is a nonstarter. We will have to capture virtually all the CO_2 and transport it to a place where it can be permanently stored. This is dirty blue hydrogen, and it can't be successful without commercial CCS systems, which include CO_2 pipelines. Such systems have largely failed in the marketplace after two decades of trying—and siting any pipeline in the United States has become difficult. So understanding the undesirability of scaling up dirty blue hydrogen also means understanding the challenges with CCS, both of which are discussed in Chapter 5.

Hydrogen from Coal

Over a quarter of all hydrogen worldwide is made with coal, the dirtiest and most carbon-intensive fossil fuel. To produce hydrogen, coal is typically gasified, impurities removed, and then its hydrogen recovered. This process results in significant CO_2 emissions, and so, from a global warming perspective, the only way such a system could be part of a hydrogen economy is with a CCS system.

The DOE has been trying to do just that for decades: Use a gasification and cleaning process that combines coal, oxygen (or air), and steam under high temperature and pressure to generate a synthesis gas (syngas) made up primarily of hydrogen and CO, without impurities such as sulfur or mercury. The water–gas shift reaction is then applied to increase hydrogen production and create a stream of CO_2 that can be removed and piped to a sequestration site. The rest of the hydrogen-rich gas is sent to a PSA system for purification and transport. The remaining gas from the PSA system can be compressed and sent to an efficient combined cycle power plant (similar to the natural gas combined cycle plants now in widespread use). Hydrogen can also power a combined cycle plant (as can syngas), so this system can be configured to generate more hydrogen and less electricity or vice versa.

As I was writing the original edition of this book in 2003, the DOE announced FutureGen (the Integrated Sequestration and Hydrogen Research Initiative), a ten-year, billion-dollar project to design and build a 275-megawatt prototype plant that would cogenerate electricity and hydrogen and sequester 90% of the CO_2 produced. The goal of the system was to "validate the engineering, economic, and environmental viability of advanced coal-based, near-zero emission technologies that by 2020"[6] will produce electricity only 10% more expensive than regular coal-generated electricity and produce hydrogen that is competitive in price with gasoline (although this did not include the cost of delivering the hydrogen). Initially, the hydrogen was to be used to generate clean power, perhaps using a solid oxide fuel cell.

But just as commercializing fuel cells has taken much longer and has proven far more difficult than was expected, so too has building large commercial coal gasification combined cycle units. One 2003 analysis described the "embarrassingly poor history" traditional power generators have had with far simpler chemical processes.[7] Aside from the unrealistic goals, the program was "badly structured, with confusion about whether it was a research project or a demonstration," as the *New York Times* summarized a 2007 Massachusetts Institute of Technology study.[8]

The project was so mismanaged and over budget it was dubbed "NeverGen." As a Congressional Research Service report explained, "Confronted with increasing projected costs in 2008, DOE under the

George W. Bush Administration first restructured FutureGen, then postponed the program when cost projections rose from $950 million to $1.8 billion."[9]

In 2010, the Obama administration restructured the moribund project as FutureGen 2.0, specifically aimed at retrofitting an existing coal plant in Illinois to capture 90% of its carbon pollution, allocating $1.1 billion in stimulus money for the effort. Obviously, it would be vastly more useful to have a CCS technology that could capture and store the CO_2 from the flue gas produced by existing coal plants than to have one that works only on newly designed plants.

But according to a 2008 US DOE report, it was not economical to retrofit existing coal plants with carbon capture technology.[10] "Existing CO_2 capture technologies are not cost-effective when considered in the context of large power plants. Economic studies indicate that carbon capture will add over 30% to the cost of electricity for new integrated gasification combined cycle (IGCC) units and over 80% to the cost of electricity if retrofitted to existing pulverized coal (PC) units," the report explained. "In addition, the net electricity produced from existing plants would be significantly reduced—often referred to as parasitic loss—since 20 to 30% of the power generated by the plant would have to be used to capture and compress the CO_2."

The 2014 Congressional Research Service report on FutureGen began with the following statement: "More than a decade after the George W. Bush Administration announced its signature clean coal power initiative—FutureGen—the program is still in early development." In early 2015, the Obama administration pulled the plug on this effort.

A decade later, the hype about this technology is back, and US companies have announced intentions to build multiple similar plants. So Chapter 5, which focuses on CCS and dirty blue hydrogen, will also examine hydrogen generation from coal plants with CCS.

Water Electrolysis Using Grid Power

Electrolyzed water is responsible for less than 1% of hydrogen. Electrolysis, the process of decomposing water into hydrogen and oxygen by using electricity, is a mature technology widely used worldwide to generate very pure hydrogen. Yet it is an extremely energy-intensive process, and the faster you want to generate hydrogen, the more power you need per kilogram produced.

Typical commercial electrolysis units use about 50–55 kilowatt-hours per kilogram, representing an energy efficiency of 70%–80%; that is, up to 1.4 units of energy must be provided to generate 1 energy unit in hydrogen. And because most electricity comes from fossil fuels, and the average fossil fuel plant is about 30% efficient, the overall system efficiency can be as low as 20% (70% times 30%); four units of energy are thrown away for every one unit of hydrogen energy produced. That is a lot of energy to waste.

"Most large-scale hydrogen production by electrolysis is using alkaline electrolysers," as one May 2024 discussion explained.[11] In the chloralkali process, electrolysis of sodium chloride (NaCl) solutions is used to make chlorine and sodium hydroxide (NaOH), also called lye, with hydrogen as a byproduct of the reaction:

$$2NaCl + 2H_2O \rightarrow Cl_2 + H_2 + 2NaOH$$

Using power from the electric grid to directly electrolyze water to generate hydrogen as an energy carrier is not a significant commercial business. Nor should it be until the electric grid is virtually carbon free, which the world is decades away from achieving.

Green Hydrogen, the Three Pillars, and Opportunity Cost

In a world with annual emissions of 50 billion tons of GHGs focused on getting as close as possible to zero in a mere quarter century, building costly and long-lived energy infrastructure that is not itself zero emission undermines efforts to avoid catastrophic climate change.

That's why green hydrogen, powered by renewables such as solar and wind, is the only form of hydrogen on which a hydrogen energy economy can be based. Unfortunately, green hydrogen is very expensive. BloombergNEF reported that it currently "costs a whopping \$4.5–\$12 per kilo."[12]

"The math still doesn't add up," the head of EU lending at the European Investment Bank told Bloomberg in May 2024.[13] "At the moment, there is no case that we can see for a model based on green hydrogen independently produced and distributed as an energy commodity."

The prospects for green hydrogen becoming affordable are not good. In the real world, the hype is giving way to the reality that the process of making green hydrogen is much more inefficient and complex than people have understood.

Green hydrogen relies on electrolyzers, which are not merely expensive pieces of inefficient equipment. They are challenging engineering projects that have recently seen sharp price increases. "The capital cost of installing an electrolysis system has increased by a median of 57%" from 2022 to 2023 in the United States, China, and Europe, *Hydrogen Insight* reported in 2024.[14] This is "rather than declining by up to 10% year-on-year," as market research firm BloombergNEF predicted in 2022.

The analysts at S&P Global Commodities reported a similar price jump in their *2024 Energy Outlook* and explained why it's happening: "The industry is encountering sizeable cost increases, with required capital expenditure estimates going up by 40–50%. While some of the cost inflation may be transitory, *electrolyzer projects tend to be highly complex, bespoke, and are proving far harder to construct than initially anticipated.*"[15] A 2023 Boston Consulting Group analysis noted that electrolyzers used to make green hydrogen production "have a cost-overrun potential exceeding 500%."[16]

In September 2024, BloombergNEF published what Martin Tengler, who heads their hydrogen research, called "the most detailed analysis of electrolysis system costs" they've ever published. It forecast capital costs falling by only "about one half from today to 2050 in China, Europe and the US assuming continued government support and free trade." If those assumptions don't prove true, cost reductions could be much smaller in Western countries. The market research firm explains this forecast "is about three times as high as what we anticipated in our 2022 analysis."[17]

Much of the hype about hydrogen is built around the mistaken belief that all energy-related technologies must continually decline exponentially in price as production increases, the way solar, wind, LEDs, and batteries have. But these technologies have achieved economies of scale, technological improvements, and overall gains in experience ("learning by doing") that translate into rapid cost reductions characterized by experience curves. In such cases, technology costs drop at a nearly constant percentage each time cumulative installed capacity doubles.

However, these are simple, modular technologies that are easy to deploy. Even so, they still took decades to become the big marketplace winners they are now. Moreover, large, complex technology projects are harder to deploy and have many components that are unlikely to see significant cost declines, including things like transformers needed for grid connection.

The authors of a 2020 study of energy systems found they could "explain systematic differences in technologies' experience rates by distinguishing between technologies based on (1) their design complexity and (2) the extent to which they need to be customized."[18] The more complicated the system design and the more they needed to be adapted and customized to their specific use environments, the slower the rate at which costs declined as sales volume increased. Solar cells are both technologically simple and easy to standardize.

But "technologies, such as nuclear power, biomass power, and carbon capture and storage" have much higher system complexity and customization for each deployment, the study noted. So they have seen the slowest rate of price decline—and, in the case of nuclear, price increases, as discussed below.

Major green hydrogen production projects are much more like nuclear and CCS than they are like solar photovoltaics or LEDs. After all, not only do you need to build an enormous complex for the electrolyzer, but you will also need a system to put the hydrogen into a form where it can be stored and transported easily. The only exception is when you use all the hydrogen as fast as you produce it at the same location, but those are very rare cases.

Transport is probably going to be in a truck carrying very high-pressure canisters, which will require a large multistage compressor system, or in a liquid hydrogen tanker, which will require a large liquefaction system to cool the hydrogen close to absolute zero and keep it at that temperature. All of this has to be done with virtually no hydrogen leakage or venting, which would otherwise wipe out whatever modest climate benefit this entire project might provide.

Any industrial facility that produces and stores large quantities of hydrogen must follow strict protocols and training procedures, generally including massive ventilation and large setback distances (sometimes 50 to 75 feet) between places where hydrogen is made and stored and other buildings (see Chapter 7). All of this adds further complexity and cost. Cutting corners can lead to serious accidents, injuries, and even deaths.

The project will be even more complex and expensive upfront if it plans to transport the hydrogen with a pipeline from the start. Such a pipeline will carry a high risk premium because it must be built before anyone knows whether the green hydrogen will be affordable or even have a guaranteed long-term market at the other end. Also, it's hard to site and build any pipeline in this country, let alone one containing a dangerous hazardous chemical that is known for being leaky and likely to embrittle any part of the pipeline system not specially designed for hydrogen.

The Three Pillars

There is another reason why major green hydrogen projects will be expensive and bespoke. Green hydrogen is truly green and carbon free only if specific rules are followed to ensure that we aren't actually just getting power from the overall electric grid with its many fossil fuel plants. The scientific literature[19] makes clear that for green hydrogen to be zero carbon, it must meet three criteria—the Three Pillars of additionality, deliverability, and hourly matching:

- Hydrogen producers must use new (additional) renewable power sources, not existing ones.
- Producers must use renewable electricity delivered from sources that directly power the nearby grid, not from distant sources.
- Producers must match *all* of their power consumption with the renewable electricity they purchase *on an hourly basis.*

These are also the proposed requirements for receiving the full tax credit from the Inflation Reduction Act for green hydrogen. First, green hydrogen must be made from renewable power plants that are new and, hence, additional. If the water is electrolyzed from existing renewables already on the grid, then you aren't actually reducing net emissions. You're just buying electricity from the grid—which still has many CO_2-emitting coal and natural gas plants—but drawing a circle around the existing renewables and pretending to claim them as yours while pretending everyone else is using the fossil fuel power. But the grid remains as dirty as it was before.

Worse, simply making hydrogen from an electrolyzer connected to the grid leads to an emission rate of CO_2 generated per kilogram "almost double the emissions rate of hydrogen produced directly from fossil fuels through steam methane reforming," explained the lead author of the 2023 study "Minimizing Emissions from Grid-Based Hydrogen Production in the United States."[20] That study found that "the marginal generation used to serve any new hydrogen load always comes in large part from fossil resources."[21]

Some companies argue that they should be allowed to claim credit for the renewables on the grid if they purchase renewable energy credits (RECs), which are certificates representing the environmental value of one megawatt-hour of renewable power. In buying a REC, you are paying a renewable project owner (or broker) a little bit of money so you can claim the entire environmental benefit of the electricity produced. Corporations and others often buy RECs to claim a smaller carbon footprint from their electricity purchases, much like a carbon offset. But as Dr. Mark Trexler explains in his 2022 online offsets course, there is "no evidence, unfortunately, the renewable energy certificate market is resulting in any more renewable energy being generated. There is no evidence of additionality."[22] Trexler calls RECS "just an accounting shell game."

The projects would've happened anyway, as many studies have found, a point also made in the 2022 *Nature* article "Renewable Energy Certificates Threaten the Integrity of Corporate Science-Based Targets."[23] This study examined "the climate change disclosures of 115 companies" with science-based targets. It found that two thirds of the claimed emission reductions from these companies' electricity usage came from RECs and so were "unlikely to be real reductions." In October 2022, *Bloomberg Green* analyzed nearly 6,000 climate reports filed by companies in 2021 and found a similar reliance on RECs in what they called "the broadest investigation yet into how companies are using this accounting technique to dramatically exaggerate their emissions reductions."[24]

Second, renewable power must be delivered from nearby sources that can power green hydrogen producers on the same electric grid. Today, such producers can claim credit for power generated far from where they use it, which means they are effectively just buying power from their local grid.

Third, green hydrogen production must match 100% of renewable power consumption *hour by hour*. Currently, a producer can purchase RECs that are averaged over an entire year. They could purchase solar power RECs that cover their annual consumption and still claim their hydrogen is made using 100% renewables, despite the reality that much of their operation occurs long after the sun has set.

Failure to enforce such rules would mean that the green hydrogen isn't green—that it has significant GHG emissions. And remember, most green hydrogen projects already have modest benefits, aren't economical, and have real opportunity costs. As noted in Chapter 1, a major 2022 study from the Electric Power Research Institute found that on average, hydrogen projects that can get the Inflation Reduction Act's 45V tax credits will achieve only one tenth of the emission benefits per dollar spent of all other clean energy projects the act funds.

Moreover, that study found that if the Three Pillars are not enforced, subsidized hydrogen projects will probably increase GHG emissions. And that's without even factoring in the inevitable increases in hydrogen leakage and venting. So we absolutely must not water down these rules. But fossil fuel companies and many other potential producers want to do just that, arguing the rules will increase the cost of green hydrogen beyond what is economical.

These rules do raise the cost of hydrogen, but they do not raise the cost of *green* hydrogen. That's because if you don't follow these rules, the hydrogen isn't green. True green hydrogen made by following all these rules isn't being sold in serious quantities now, and it won't be cheap any time soon.

The opponents of the Three Pillars argue that requiring true green hydrogen would so expensive it would delay the emergence of a hydrogen economy. They say they should be allowed to get huge tax breaks to make nongreen hydrogen for several years so as to start establishing the entire infrastructure of a hydrogen economy without waiting for green hydrogen to become competitive.

Yet that argument assumes the hydrogen economy is inevitable, so anything we do to speed the transition is a good idea even if it increases carbon pollution in the short term. But as this book makes clear, so many intractable obstacles to a hydrogen economy remain that major government policy should not be based on any claim of inevitability for green hydrogen as a major energy carrier.

The DOE has an incredibly optimistic projection of an 80% price drop—from $5 down to $1—for the price per kilogram of green hydrogen by 2030 if we scale up production. But as discussed, bespoke technology systems such as electrolyzer-produced hydrogen with onsite storage and large renewable energy projects nearby do not inevitably keep dropping in price.

The 45V tax credit for green hydrogen—a remarkable $3 per kilogram for genuinely very low-carbon hydrogen—is so large it will distort the market and lead to investments that don't make sense from either an economic or climate perspective. Even worse, the tax credit does not decline over time even though it can be used for ten years from when production starts for any system that begins construction by the end of 2032.

What happens if the production cost does drop significantly over the next decade? Jesse Jenkins, a Princeton professor who has worked on analysis that demonstrates the dangers of bad hydrogen policies, told *Canary Media* in 2024, "I do worry that by the end of that period, it will make sense simply to produce hydrogen and flare it or make a subsidy farm."[25] That is, it might make economic sense to produce hydrogen and burn it on site for no productive purpose.

There's one more huge problem with the hundreds of billions of dollars of hydrogen hype. Truly green hydrogen is certainly the only hydrogen worth supporting by the government or using by industry, but scaling it up is still a terrible idea for at least the next two decades because of the huge opportunity cost.

Opportunity Cost

Under what circumstances would green hydrogen be a good use of carbon-free power? There is an opportunity cost to using vast amounts of renewables (or nuclear power) in an expensive and inefficient effort to reduce CO_2 by trying to make hydrogen work as an energy carrier. That carbon-free power could have been used to directly replace the CO_2 emissions from fossil fuel plants—and from vehicles—much more cheaply and efficiently.

For example, hydrogen vehicles are a very inefficient way to use carbon-free energy. Converting renewable energy to green hydrogen with an electrolyzer—then storing, transporting, and pumping it into a vehicle's onboard fuel cell just to regenerate electricity to power an electric motor—means throwing away as much as 80% of the original renewables. And it relies on expensive electrolyzers and fuel cells.

But that complex process clearly can't compete with simply using the original renewable electricity to charge the battery on an electric vehicle and then running the electric motor off the battery, which loses less than 20% of the renewables. So you can go four times further on the same renewable power with an electric vehicle than with a hydrogen car, at a far lower cost.

The same applies to other key sectors when electricity goes head to head with hydrogen. Independent studies reveal that five to six times more electricity is needed to generate the same amount of heat with hydrogen than with an efficient heat pump.[26] Other highly inefficient uses of renewable electricity, according to a 2023 analysis, are making green hydrogen to directly displace natural gas power generation and "blending hydrogen into the gas grid."[27]

This big opportunity cost is too often absent from media and policy discussions. If we misallocate all those renewables, we lose the opportunity to achieve far greater reductions for far less money by using them to replace fossil fuels directly. At the rate the nation and the world have been cutting emissions—not fast enough and not at all, respectively—green hydrogen won't make sense until after 2050.

Hydrogen from Biomass

Biomass can also be gasified and converted into hydrogen (and electricity) in a process very similar to the integrated gasification combined cycle used to convert coal. Biomass is any material that has participated in the growing cycle. It includes agricultural, food, and wood waste, as well as trees and grasses grown as dedicated energy crops. Biomass is of interest from a global warming perspective because—in theory—when it is burned or otherwise transformed into energy, it releases CO_2 previously sequestered from the atmosphere (temporarily) in the growing cycle, so the net CO_2 emitted could be zero.

But there are two problems with making hydrogen from biomass. First, biomass gasification remains a problematic technology. In the 2004 first edition of this book, I wrote, "A number of biomass gasification processes are being demonstrated, although a practical commercial system remains technologically challenging." Two decades later, those technological challenges have not been overcome. In November 2022, Dr. Zach El Zahab, Gasification Program manager for GTI Energy (formerly the Gas Technology Institute), discussed ten technical challenges to large-scale commercial biomass gasification at a DOE gasification technology workshop. "Biogenic feedstocks gasification continues to face several technical challenges," he concluded. "All solutions need to be proven and matured on a commercial scale."[28]

Indeed, as a major December 2023 report on carbon removal from leading experts explained, "While smaller proof-of-concept projects have been shown, full-scale biomass gasification projects are yet to be demonstrated."[29] The report, which was led by Lawrence Livermore National Laboratory with DOE support, added, "The largest biomass gasification biorefinery in the world is in Finland and only processes about 526 metric tons per day of forestry biomass." That is remarkably little progress in two decades.

Second, in the real world, scaling up bioenergy is not a sustainable way of providing energy without accelerating global warming. That is clear from original modeling in 2023 from the researchers at Climate Interactive and the Massachusetts Institute of Technology, whom I worked with, and the scientific literature of the last few years, all of which is discussed in the second half of Chapter 6.

The Hype About Natural Hydrogen

The newest driver of hydrogen hype is the possibility that natural or geologic hydrogen found underground and generated through natural chemical reactions might play a major role in fostering a hydrogen economy. In reality, as we've discussed, hydrogen is a lousy energy carrier with several fatal limitations, even if we solve the problem of producing low-cost carbon-free hydrogen. Currently, there is no reason to believe that natural hydrogen will solve that problem, as made clear by work from the Hydrogen Science Coalition, "a group of academics, scientists and engineers seeking to bring an evidence-based view to hydrogen's role in the energy transition."[30]

The coalition wrote in a March 2024 article, "The most promising geologic hydrogen finds to date lack decarbonization potential"; there is just too little, it's generally in the wrong place, and often there isn't actually much hydrogen in the gas that has been found. As a result, "the odds of finding geologic hydrogen that can be extracted at the scale of large natural gas developments look relatively slim."[31]

Yet we see the same hydrogen hype from the normally more skeptical scientific and financial media—and even, unfortunately, scientific journals.

"Geologists Signal Start of Hydrogen Energy 'Gold Rush,'" read the February 2024 headline from the *Financial Times*.[32] The article points to a site in Mali where "since 2012, almost pure hydrogen has flowed from a borehole there with no diminution of pressure." The newspaper says it "may have inspired a hydrogen rush comparable with the birth of the petroleum industry in 1859, when Edwin Drake drove a pipe into the ground at Titusville, Pennsylvania and struck oil."

It would have been useful to tell readers that Mali has the world's only known hydrogen producer well—putting out just 0.00005% of current world hydrogen demand—the power output equivalent to "less than a tenth of a single medium-sized wind turbine," as the Hydrogen Science Coalition noted.

Also, there is no analogy between oil and hydrogen. Oil has one of the highest known energy densities of any fuel, and it's easy to store, transport, and put onboard a vehicle. By contrast, hydrogen has the lowest energy per unit volume of any potential energy carrier and is perhaps the hardest to store, transport, and put onboard a vehicle. When compared with oil, hydrogen is also very leaky, as Stuart Haszeldine, a University of Edinburgh geologist, told *Yale Environment 360*. "It leaks almost as fast as it is produced, and certainly over geological time, it's not accumulated to any great extent."[33]

Because hydrogen is so hard to store and transport, the best use of whatever usable geologic carbon is found would probably be to run an inexpensive electric generator, which is precisely what those living around the Mali site do.

But we have seen a gold rush—for startups. In a March 2024 news release, "The White Gold Rush and the Pursuit of Natural Hydrogen," research firm Rystad Energy reported that the number of companies looking for natural hydrogen had jumped from ten in 2020 to forty by the end of 2023.[34] In the pointless and confusing color scheme, natural hydrogen is called either white or gold hydrogen.

The biggest winner to date is the startup Koloma, which, as Canary Media put it in February 2024, "just raised an eye-popping, gobsmacking $245 million round of venture funding," led by Khosla Ventures and including Bill Gates's Breakthrough Energy Ventures, Amazon's climate fund, and United Airlines' venture capital arm.[35]

The only gusher natural hydrogen has seen has been a gusher of overhyped media, especially in 2024. *Semafor* ran an article on geologic hydrogen asserting, "Hydrogen is a kind of magic key in the energy transition,"[36] and *The Colorado Sun* ran the headline, "The hunt is on for Colorado's next gold rush: Geological hydrogen."[37] CNBC's piece was headlined, "A global gold rush for buried hydrogen is underway—as hype builds over its clean energy potential."[38]

Hype indeed. In contrast, the International Energy Agency in its 176-page *Global Hydrogen Review* 2023, notes, "Extracting natural hydrogen requires compression and purification systems since natural hydrogen accumulations contain different impurities (such as nitrogen, methane, CO_2, helium and argon) and the hydrogen content can vary widely (2.4–100% in volume)."[39] If the hydrogen comes with significant amounts of methane or CO_2, it can be hard to extract and potentially have very high GHG emissions. The International Energy Agency also cautions, "There is a possibility that the resource is too scattered to be captured in a way that is economically viable." It may be a dead end.

A China-based multinational think tank wrote in May 2024, "Commercial production of white hydrogen is unlikely to be achieved in the short term, and it is even less likely to significantly change the current situation of hydrogen energy and even the new energy supply."[40]

The hype about natural hydrogen reached a frenzy in early 2024 thanks to the journal *Science*. First, on December 14, 2023, *Science* named the accelerating "hunt for natural hydrogen" a runner-up for the 2023 "Breakthrough of the Year."[41] Then it published a study in February 2024 on a geologic hydrogen discovery in Albania.[42] This was accompanied by a news article, "Gusher of Gas Deep in Mine Stokes Interest in Natural Hydrogen."[43] The article said, "One takeaway from the discovery" is that natural "hydrogen seems to be more ubiquitous than once imagined." How ubiquitous? The journal quotes an energy system researcher saying, "It could really disrupt geopolitics."

What was this world-changing discovery? The study abstract stated, "We report direct measurements of an elevated outgassing rate of 84% (by volume) of H_2 from the deep underground Bulqizë chromite mine in Albania. A minimum of 200 tons of H_2 is vented annually from the mine's galleries, making it one of the largest recorded H_2 flow rates." It ends by saying, "This discovery suggests that certain ophiolites may host economically useful accumulations of H_2 gas." Ophiolites are rock formations containing iron-rich rocks that can generate hydrogen when they react with water at high temperatures and pressure.

An annual flow rate of 200 tons equates to 0.0002% of global hydrogen production. Also, you might be forgiven for thinking that the 200 tons of hydrogen were all outgassing as a mixture of gases that were 84% by volume hydrogen. But you must read the entire study to learn that only 11 tons of hydrogen per year—0.000011% of global production—outgasses at that potentially useful density. And that outgassing is 13% methane, a potent heat-trapping gas. So the carbon intensity of hydrogen commercially produced from that mixture could be high, according to one 2023 study.[44]

The rest of the outgassing is a 1.2% hydrogen mixture flowing at 42 tons per year of hydrogen and a 0.4% mixture flowing at 158 tons per year. It is very unlikely that hydrogen can be produced commercially from such a diffuse outgassing economically and with a low carbon intensity.

The accompanying news article waits until the end to inform us that the researchers estimate that the site might contain as little as 5,000 tons of hydrogen and that another expert doubts this can be commercially exploited. Indeed, the article then points out, "Prospectors should be aiming for hydrogen deposits of 10 million tons or more, according to a recent natural hydrogen grant funding announcement" from DOE. So this gold mine could be 2,000 times too small to be practical. Again, even that is optimistic because of how diffusely 95% of the hydrogen is outgassing.

Given how the study authors and *Science* itself hyped up this story, it's no surprise that it helped inspire *National Geographic*'s March 2024 article, "The Gas in This Exploding Mine Is Odorless, Colorless—and Could Transform the World."[45] In May 2024, *Scientific American* ran a similar-sounding headline, "'Hydrogen Fever' Erupts After Discoveries of Large Deposits of the Clean Gas."[46] Ironically, you might think that "fever" is being used in its slang sense, "a state of great excitement or enthusiasm."[47] But the magazine is actually quoting the lead author of the study trying to tamp down the hype: "Truche casts a cautionary note, calling the fervor over the discoveries a 'hydrogen fever.' 'In one sense, it is wonderful because it attracts attention, funding and lots of motivation to move forward,' he adds. *'But in another sense, it's also kind of a Wild West—with lots of overstatements.'*"

Lots of overstatements.

Canary Media quotes a partner in one of the venture capitalists backing Koloma, who calls geologic hydrogen the "most environmentally friendly way of making hydrogen that's out there." The article calls that "a conclusion supported by early peer-reviewed research on the subject." That research, also quoted by *Scientific American* and many others, comes from a study published by *Joule* in August 2023 that was funded by Smart Gas Sciences, "a company interested in geologic hydrogen."[48]

The study makes some out-of-date assumptions that push the final calculation toward an unrealistically low GHG intensity. A more up-to-date analysis would reveal that natural hydrogen is not environmentally friendly, especially over the next two decades, when we must slow warming as fast as possible to avoid crossing dangerous temperature thresholds that lead to far more catastrophic climate change.

The 2023 *Joule* study downplays the recent upward revisions in the global warming potential (GWP), so they are not used in the base case. The study also focuses on the old estimate of warming impact over 100 years (GWP100). Yet the world has now prioritized reducing so-called short-lived climate forcers such as methane (the main constituent of natural gas), whose warming impact over twenty years (GWP20) is eighty times that of CO_2.

Significantly, the revised GWP20 of hydrogen is seven times higher than the old GWP100.[49] Also, methane is often found mixed in with natural hydrogen. The hydrogen outgassing in Albania is 13% methane, a level that sharply raises the GHG intensity of hydrogen production. If you then use the GWP20 of both hydrogen and methane, rather than the GWP100, this would be quite dirty hydrogen.

The *Joule* study uses as its base case an 85% hydrogen mixture with little methane. But as the Hydrogen Science Coalition noted in March 2024, although commercial exploitation requires at least 60% of the sampled gas to be hydrogen, the actual content "is relatively low in many hydrogen finds to date—often under 40%."[50] Indeed, we saw in the much-hyped Albania case that 95% of the hydrogen came out in a very diffuse mixture, with most just 0.4%.

Also, the study points to a 2020 review article on natural hydrogen, which documents that a majority of concentrations measured in hydrogen finds going back decades are lower than 60%. The review also notes that in one study of a well drilled in Russia, "a flow of hydrogen gas with up to 87.3%" was encountered at a depth of a half mile. It "continued to flow for several months, and *hydrogen concentrations progressively declined to 0.5%, being totally replaced by* CO_2."

Finally, the *Joule* article excluded "cases that would likely have high climate impact and therefore are unlikely to be implemented in a clean energy future (e.g., a case where entrained methane is separated then vented)." But in the real world, people can and do make precisely such money-saving shortcuts, as we've learned from countless media exposés, lawsuits, and studies on the voluntary carbon market because it is unregulated and there is no way to enforce standards.[51]

The bottom line is that it is quite premature to assume that geologic hydrogen will have a low carbon intensity. Right now, it looks like it will probably be much higher than green hydrogen.

Hydrogen from Nuclear Plants

Nuclear power plants can also be a source of hydrogen. Electrolysis of water with electricity from a nuclear power plant is possible, but that doesn't make sense for the foreseeable future, probably through at least 2050, either economically or environmentally. It doesn't make sense for big nuclear plants, and it makes less sense for smaller reactors. This over-hyped and overfunded means of hydrogen production will be the focus of Chapter 4.

CHAPTER 4

The Hype About Hydrogen from Nuclear Power

Nᴇᴡ ɴᴜᴄʟᴇᴀʀ ᴘᴏᴡᴇʀ ᴘʟᴀɴᴛꜱ ʜᴀᴠᴇ become steadily more expensive for so long that the United States and Europe have all but stopped building any. Global nuclear power production was lower in 2022 and 2023 than it was in the mid-2000s.[1] As a share of global power production, nuclear energy peaked just above 17% in the mid-1990s, but by 2022 and 2023 it was down to 9%.

It's not uncommon for new nuclear power projects to "witness up to a staggering 400% in overruns," a 2023 Boston Consulting Group study noted.[2] The only new nuclear plant the United States built in recent decades, the Vogtle plant in Georgia, was wildly over budget and cost $35 billion. It is "the most expensive power plant ever built on earth," explained *Power Magazine* in August 2023, with an estimated electricity cost "which is astoundingly high."[3]

The first new small nuclear reactor the United States tried to build— by NuScale—was canceled in 2023, after its cost per megawatt was projected to exceed Vogtle's. In June 2024, Bill Gates told CBS News that the full cost of his new small Natrium reactor under construction in

Wyoming, would be "close to $10 billion,"[4] meaning it has a shocking cost per megawatt twice that of Vogtle.

Such an outcome was not surprising. Indeed, it was predicted by the *IEEE Spectrum* in a 2015 article, "The Forgotten History of Small Nuclear Reactors: Economics Killed Small Nuclear Power Plants in the Past—And Probably Will Keep Doing So."[5] A 2024 analysis of proposed small reactors found none "are fit for necessary rapid decarbonization due to availability constraints and economic challenges."[6]

Yet worldwide plans for hydrogen from nuclear power have exploded. In November 2022, the Department of Energy (DOE) announced, "3 Nuclear Power Plants Gearing Up for Clean Hydrogen Production."[7] In September 2023, Reuters reported, "European lawmakers bowed to pressure from France to allow nuclear power to be used for ammonia and hydrogen production" so they could enact new legally binding EU targets for expanding renewable energy.[8] The same month, Westinghouse announced it was "exploring" using electrolysis of water to make hydrogen "at existing Light Water Reactors and in our advanced reactor designs."[9]

Also in September 2023, Reuters quoted a senior DOE official who oversees billions of dollars in loans for clean energy who said, "The whole concept of nuclear and hydrogen is one that makes a lot of intellectual sense."[10] The next month, *Power* magazine reported that Constellation Energy intended to use some of the federal funding for the Midwest Hydrogen Hub—one of seven such billion-dollar hydrogen production centers planned around the country—to build "the world's largest nuclear-powered clean hydrogen facility."[11] The month after that, the government-owned Nuclear Power Corporation of India said it was setting up hydrogen generation at two of its plant sites.[12]

In April 2024, *Nikkei Asia* reported that the Japanese government intended to begin "field testing clean hydrogen production using nuclear as soon as 2028."[13] The next month, *Nucnet* reported that a Swedish nuclear power company had agreed to deliver hydrogen to Hynion, whose focus is hydrogen refueling stations.[14] Then, in June, South Korea announced that several companies, including Hyundai and Samsung, would build their first nuclear hydrogen project by 2027.[15]

But the reality remains that none of these plans make sense from an economic, environmental, or safety perspective—and that will be true for at least the next two decades and probably much longer. Nuclear power plant economics became so bad this century that it wasn't only *new* nuclear plants that were uneconomical. By 2017, the low cost of natural gas and the declining costs for renewable power and energy efficiency created a situation where over half of *existing* US nuclear power plants were no longer profitable to run. Bloomberg New Energy Finance (now BloombergNEF) calculated in a 2017 analysis that thirty-four of the nation's sixty-one nuclear plants "are bleeding cash, racking up losses totaling about $2.9 billion a year."[16] If simply operating most existing nuclear power plants was unprofitable, then it's no surprise that actually building and operating an expensive new plant was wildly uneconomical.

As a result, many people started hyping advanced small modular reactors (SMRs) as a way to make nuclear cost-effective, but they have as long a history of failure as bigger nuclear plants. Also, smaller plants mean abandoning the enormous economies of scale that drove nuclear plants to get big everywhere in the world. So they will probably have as high or higher a cost per megawatt to construct as the most expensive big plants do.

Indeed, it would be unprecedented in the history of energy for these smaller reactors to overcome not only the high cost of nuclear plants but also the diseconomies of shrinking them down and to somehow drop in price so sharply that SMRs become such clear marketplace winners as to make a major contribution to cutting CO_2 emissions by 2050. Studies have found that standardizing design can have a modest economic benefit but "that innovation in nuclear power increases construction costs," a result that "contrasts with the pattern seen in other energy technologies, where technical progress contributes to costs reductions, notably in other competing low carbon technologies."[17]

From a safety perspective, hydrogen is a uniquely hazardous chemical, one that the nuclear industry has studied and worried about for decades. The scientific literature is clear that putting a system to generate hydrogen close to a nuclear reactor is very risky.[18]

From a climate perspective, using nuclear power to produce hydrogen makes even less sense than using renewable power. As we have seen, you can achieve far more emission reductions far more cheaply by using renewable power to replace fossil fuels directly. And that will be true until we have eliminated most CO_2 emissions in the economy, at least 90%. The same analysis applies to nuclear power: Compared with using it to make hydrogen, nuclear power can reduce far more greenhouse gases far more cheaply by directly replacing fossil fuels.

But there's one big difference. Nuclear already serves a purpose for which advocates say hydrogen might be needed: It can provide electric power when the sun isn't shining and the wind isn't blowing. Taking this expensive but useful form of carbon-free power and throwing most of it away to make hydrogen, store it, and then convert it back to electricity when renewables aren't available makes little sense. That's especially true with all the advances in long-term energy storage we now are now witnessing. Indeed, a June 2024 study concluded that "dedicated production of clean hydrogen via electrolysis for use as a power plant fuel is unlikely to yield an effective decarbonization outcome."[19] As expensive as it is, simply deploying nuclear power directly would "likely be more cost-effective" for supporting the grid than for using it to make hydrogen for the same purpose.

A (Brief) History of Nuclear Power Plants

Nuclear power may be the original overhyped energy technology, as an article on the US Nuclear Regulatory Commission (NRC) website makes clear.[20] In a 1954 address to science writers, Atomic Energy Commission chairman Lewis Strauss said, "Transmutation of the elements, *unlimited power*, ability to investigate the working of living cells by tracer atoms, the secret of photosynthesis about to be uncovered—*these and a host of other results all in 15 short years. It is not too much to expect that our children will enjoy in their homes electrical energy too cheap to meter.*"[21]

Strauss repeated the idea just days later on a *Meet the Press* radio broadcast, saying that he expected his children and grandchildren to have power "too cheap to be metered." That time, he said, may be "close at hand. I hope to live to see it."

The United States did develop a nuclear industry, and we ultimately built over 100 reactors, more than any other country. But the industry did not see reactor prices going down an experience curve, where increased sales over time lead to economies of scale, improvements in technology, and overall gains in experience that translate into steady cost reductions—as they have in recent decades with solar energy, wind power, batteries, and LED bulbs. Instead, new nuclear power plants have steadily risen in price. This "negative learning" as one article called it, happened in both the United States and France.[22]

As a result, nuclear power has largely priced itself out of the market in the industrialized world. "Western nuclear completions since 1990 took many years and resulted in massive cost overruns," as JP Morgan explained in a 2024 analysis.[23] "We estimate that levelized nuclear costs were 2x–4x higher than a baseload power system derived from wind, solar and sufficient backup thermal (natural gas) capacity."

I have been involved with nuclear energy policy and analysis for over thirty years. When I first came to the DOE in mid-1993, I spent two years as special assistant for policy and planning for the deputy secretary, who oversaw all DOE energy programs. One of my duties was to review policy and analysis coming from the Office of Nuclear Energy.

Over the years, I have researched and written about nuclear power for studies, books, and countless articles. In the 2004 edition of this book, I noted that a major 2003 interdisciplinary study by the Massachusetts Institute of Technology, *The Future of Nuclear Power*, highlighted many of the "unresolved problems" that have created "limited prospects for nuclear power today." The study found that "in deregulated markets, nuclear power is not now cost competitive with coal and natural gas." The challenge of siting new nuclear power plants is exacerbated by public concern about the safety, environmental, health, and terrorism risks associated with nuclear power. The study found that "nuclear power has unresolved challenges in long-term management of radioactive wastes." It described possible technological and other strategies for addressing these issues but noted, for instance, that "the cost improvements we project are plausible but unproven."[24]

Such improvements never happened. In 2008, I was invited to testify on the economics of nuclear power by the Senate Committee on Environment and Public Works Subcommittee on Clean Air and Nuclear Safety.[25] As I testified, the cost of new nuclear power had more than doubled from what the Massachusetts Institute of Technology report assumed in its base case just five years earlier. From 2000 through 2007, nuclear plant construction costs—mainly materials, labor, and engineering—rose by 185%.[26] That meant a nuclear power plant costing $4 billion to build in 2000 cost over $11 billion to build seven years later. An industry trade magazine, *Nuclear Engineering International*, titled a 2007 article "How Much? For Some Utilities, the Capital Costs of a New Nuclear Power Plant Are Prohibitive."[27]

The only new nuclear power plants the United States successfully built and started in recent decades are Units 3 and 4 of the Vogtle plant, operated by the Southern Company and its subsidiary, Georgia Power. A 2006 *New York Times* article posing the question "A Nuclear Renaissance?" reported that Westinghouse told the paper, "The cost will ultimately be somewhere between $1.4 billion and $1.9 billion" for each AP1000 reactor.[28] Yet the *Wall Street Journal* reported two years later that "the existing Vogtle plant [Units 1 and 2], put into service in the late 1980s, cost more than 10 times its original estimate, roughly $4.5 billion for each of two reactors."[29] The same article suggested the two planned units would cost $14 billion total.

Ironically, that *Journal* article hyped the supposed nuclear Renaissance, asserting, "Nuclear power is regaining favor as an alternative to other sources of power generation, such as coal-fired plants." But that part of the story was inaccurate, as the few nuclear plants then under consideration were canceled one by one until only the two Georgia reactors were left. By the time they were turned on, seven years late, one in 2023 and one in 2024, their total cost had hit $35 billion.[30]

And that isn't just the US experience. France's government-owned electric company, EDF, has had the same outcome with the 1,600-megawatt European pressurized-water reactor (EPR) Generation III+ reactor design developed with Germany's Siemens. As of 2024, the only reactor project currently being constructed in France was a single EPR plant at Flamanville. The original cost estimate was €3.3 billion. The current cost estimate is nearly six times as high, €19.1 billion ($19 billion).[31] Similarly, the Olkiluoto nuclear plant in Finland "was scheduled to be completed in 2009; it was completed in 2023 and cost $12 billion, three times its original estimate," as JP Morgan noted in 2024.[32]

In a 2008 "White Paper on Nuclear Power," the British government's

Department for Business, Enterprise & Regulatory Reform estimated a "total cost of £2.8 bn to build a first of a kind plant with a capacity of 1.6 GW" for a single reactor.[33] That analysis asserted, "Even on cautious assumptions, the cost of nuclear energy compares favourably with other low-carbon electricity sources."

Again, this was more empty hype. The country pursued two EPRs, 3,200 megawatts total, at the Hinkley site in southwest England. This plant would have the country's first two new reactors since the 1990s. In January 2024, the BBC reported, "EDF now estimates that the cost could hit £46bn" ($59 billion).[34] That is a price per reactor eight times higher than the 2008 report had projected. The start date was pushed back to at least 2029. China General Nuclear Power Corp, which owns about a third of the project, with EDF owning the rest, halted funding in December 2023, and EDF has warned the halt could become permanent.[35]

"It seems the golden rule of nuclear economics is to add a zero to industry estimates, and your estimate will be far closer to the mark than theirs," notes nonprofit news service Climate & Capital Media in a January 2024 report.[36]

The Hype About SMRs

More than eight decades—and trillions of dollars—after the world's first artificial nuclear reactor was built, new large nuclear reactors are clearly not an affordable, scalable solution to climate change, nor is it plausible to think they will be any time soon. And that means they're not a plausible way to reduce carbon-free hydrogen prices to affordable levels.

So, in one of the great rebranding moves, the nuclear industry and its supporters have been promoting SMRs—those 300 megawatts or smaller. They've pushed the idea that SMRs are a new and qualitatively better type of reactor that we should believe has a very real chance of solving all the problems that made large nuclear plants figuratively radioactive to utilities: "poor economics, the possibility of catastrophic accidents, radioactive waste production, and linkage to nuclear weapon proliferation."[37]

But that's all hype. As many studies and experts make clear, SMRs aren't new, they aren't qualitatively better, and there is little chance they could solve all four problems. For instance, the lead author of a 2022 Stanford-led study on the waste question explained, "Our results show that most small modular reactor designs will actually increase the volume of nuclear waste in need of management and disposal, by factors of 2 to 30" for reactors they analyzed.[38]

The study warns that SMRs are "incompatible with current technologies and concepts for nuclear waste disposal."[39] As a result, SMR waste will need special treatment, conditioning, and packaging: "These processes will introduce significant costs—and likely, radiation exposure and fissile material proliferation pathways—to the back end of the nuclear fuel cycle and entail no apparent benefit for long-term safety."

So SMRs may have higher cost, higher waste, the same or worse proliferation risk, and the same or worse safety. Indeed, a 2014 article in the journal *Energy Research & Social Science* argued that trying to solve the economic problem will probably make at least one of the others worse.[40]

Consider NuScale, featured in a Harvard Business School case study from 2014, "NuScale Power—the Future of Small Modular Reactors."[41] Harvard called NuScale "the leading modular nuclear reactor in the United States. This Reactor will be the safest and simplest ever built." In September 2020, *Popular Mechanics* asserted, "This Tiny Nuclear Reactor Will Change Energy."[42]

The NRC accepted NuScale's design certification application in March 2018 and certified the design in July 2022, making it the first SMR approved by the NRC for use in this country. In December 2022, NuScale and Shell Global Solutions announced a plan to develop a "system for hydrogen production using electricity and process heat" with partners, including the DOE's Idaho National Laboratory and Utah Associated Municipal Power Systems (UAMPS).

NuScale, the Idaho lab, and UAMPS were at the time also working to deploy six 77-megawatt NuScale modules near the lab. But in November 2023 NuScale and UAMPS canceled that 462-megawatt project, which was to be the first US SMR project, after the projected price of electricity surged over 50% in just eighteen months. On a conference call explaining the decision, NuScale CEO John Hopkins said, "Once you're on a dead horse, you dismount quickly. That's where we are here."[43]

That surge was "due to a 75% increase in the estimated construction cost for the project, from $5.3 to $9.3 billion," according to a report from the Institute for Energy Economics and Financial Analysis. This meant the projected cost per kilowatt to build the NuScale SMR, over $20,000 per kilowatt-hour, was higher than the cost to build the two latest Vogtle units, about $15,000 per kilowatt-hour.[44]

NuScale's projected price of electricity had jumped sharply to $89 per megawatt-hour, or 8.9 cents per kilowatt-hour. "That's more expensive than most other sources of electricity today, including solar and wind power and most natural-gas plants," *Technology Review* explained in February 2024.[45] Also, the projected electricity price would have been far higher if not for the massive taxpayer subsidy, such as the $1.4 billion DOE kicked in and the 2022 Inflation Reduction Act, with its nuclear power credit of $30 per megawatt-hour, or 3 cents per kilowatt-hour.

Moreover, there's every reason to believe NuScale's delays and cost overruns would have continued to escalate through construction because that's the overwhelming historical trend in the United States and Europe. We've already seen that with the big reactors, but it's also true with the smaller ones. In Russia, for example, two SMRs began commercial operation in May 2020 after significant delays and cost overruns.[46]

In 2021, the reactors' load factors were only 45% and 18%, respectively. Load factor is how much power a reactor actually delivers compared with what it would deliver running at maximum power.

China is often held up as a country that doesn't have the same challenges as the United States in building nuclear plants, and indeed they are now building many large plants. But the problems with SMRs are universal. China's first SMR was connected to the grid in December 2021, a 105-megawatt high-temperature gas-cooled reactor pebble-bed module. *The World Nuclear Industry Status Report 2022* explained, "Delays and cost escalation in this project offer an excellent illustration of why SMRs are likely to be no different from reactors with higher power ratings."

Originally, the CEO of the joint venture between the state-owned China Nuclear Engineering Group and Tsinghua University's Institute for Nuclear and New Energy Technology said construction would start in spring 2007, with operation beginning "by the end of the decade." When construction actually started in 2012, the estimated completion time had increased to fifty months. It actually took twelve years.

In 2021, the Chinese started building a second SMR design: the 125-megawatt ACP100 reactor. A Chinese National Nuclear Corporation official said construction would take nearly six years. As the *World Nuclear Industry Status Report 2022* explained, "by the time construction started in 2021, this SMR was at least six years late," and "the reactor will also not be economical." The Chinese National Nuclear Corporation admitted in 2021 that the cost per kilowatt of the proposed ACP100 demonstration project "is 2 times higher than that of a large" nuclear power plant, and the cost per kilowatt-hour is likely to be 50% higher.

The delays and rising costs of SMRs worldwide should not have been a surprise. The history of nuclear power reveals the repeated failure of commercial SMRs to prove practical or affordable and an endless push to capture economies of scale, as the *IEEE Spectrum*, the leading publication of the Institute of Electrical and Electronics Engineers, made clear in a 2015 article.[47]

Most of the expenses of building and running a nuclear plant do not increase directly in proportion to the power it generates. Building a 300-megawatt reactor doesn't require half as much steel and concrete as a 600-megawatt reactor. It requires *more* than half. And it requires more than half as many people to run.

According to the standard "power rule" used in industries such as nuclear for the capital cost of production facilities, a 300-megawatt plant would have nearly twice the cost per megawatt of capacity as a 1,000-megawatt plant. The reverse of economies of scale is diseconomies of shrinkage.

It is no surprise, then, that "the pursuit of economic competitiveness drove the attempt to reap economies of scale, resulting in larger unit sizes," as an April 2024 "techno-historical" analysis documented.[48] The average electric capacity of nuclear reactors worldwide, which was below 300 megawatts from the mid-1950s to the mid-1960s, rose to nearly 1,000 megawatts by the mid-1980s. Over the next two decades, the capacity fluctuated downward to under 800 megawatts but then started climbing again to nearly 1,200 megawatts in the early 2020s. This recent rise occurred as new reactor builds all but stopped in the United States and Europe, while the big new nuclear plant builders, like China, all saw the benefit of economies of scale.

Significantly, not only were the plants getting bigger, but even the bigger plants were sited together. As a 2018 Nuclear Energy Institute report noted, "approximately 80% of the electricity generated from nuclear power in the U.S. comes from plants with multiple reactors."[49]

A major reason for this is that one of the biggest costs and delays with any proposed nuclear plant is getting every necessary approval from the various constituencies in a state or in the local community that can delay or block siting and construction. After all, a great many people do not want to live or work near a major nuclear power plant. So as the power plant manufacturer and utility go through this lengthy and onerous process, they naturally want to cram as much power into the site as possible. This is another economy of scale that drives power plants to be so big.

The possibility that siting smaller nuclear plants is somehow going to be much faster and smoother has not been seen historically. And that's why we already see multiple SMRs typically sited together, as was the case with NuScale. But that raises a question: If most of the applications for an SMR are going to involve multiple units in the same place, the manufacturer is going to be driven toward simply building bigger plants, which is exactly what has happened over the past seventy years.

One of the strangest aspects of the SMR discussion is that more than fifty SMR designs are currently in various stages of development. But the only rationale for SMRs is the possibility that they will overcome the inherent diseconomies of shrinkage by achieving some sort of "economies of standardization"—the gains you might achieve if everyone could settle on one or at most two designs. After all, one of the big failures of the US nuclear industry was the inability to agree on a single design that could make licensing, siting, and construction simpler and potentially less expensive. But if the United States were to have, say, four or five competing SMR designs, then it seems unlikely that any of those would achieve economies of standardization because the market in this country (and Europe) for new nuclear plants of any size is not huge. So the overwhelming majority of startups built around SMR designs seem destined to fail.

Remember, there is no reason to believe that the economies of standardization, if they actually do manifest, would be large enough to overcome both the diseconomies of shrinkage and the inherently high cost of nuclear plants. But that is exactly what would be needed to create a successful commercial SMR to compete in the market of the 2030s and beyond. That's especially true with all the advances we are seeing in competitors to nuclear power, including long-duration energy storage and enhanced geothermal energy systems.

A 2014 journal article concluded, "We argue that scientists and technologists associated with the nuclear industry are building support for small modular reactors" by putting forward "rhetorical visions imbued with elements of fantasy that cater to various social expectations."[50] These include the possibility of "risk-free" energy by using terms such as "passive safety" and "inherently safe." They warn that these visions involve "downplaying," or "erasing" and "selectively presenting," data about "the cost and economic competitiveness of SMRs."

We've already seen that the entire discussion about nuclear power, SMRs, and hydrogen from nuclear power involves downplaying, erasing, or selectively presenting data about the cost and economic competitiveness. The same fantasy applies on the safety side.

Returning to NuScale, physicist and nuclear safety expert Dr. Edwin Lyman noted that while developing the reactor, "NuScale made several ill-advised design choices in an attempt to control the cost of its reactor, but which raised numerous safety concerns."[51] For instance, "the design lacked leak-tight containment structures and highly reliable backup safety systems." Some of the money-saving choices were justified on the basis that the reactor was "passively safe," but one of the NRC's own experts raised serious questions about the passive emergency core cooling system late in the design certification process.[52] Similarly, two other leading experts cast doubt on the reactor's safety and the NRC's certification process.[53] Yet even with all this apparent skimping on safety, the reactor turned out to be unaffordable.

The scant attention to safety in the public debate about hydrogen from nuclear power is particularly worrisome because it appears to reflect a complete lack of awareness of just how hazardous a fuel hydrogen is.

Thermoelectric Hydrogen Production and Nuclear Safety

As we've seen, producing hydrogen by using electricity from a nuclear plant does not make much sense economically or environmentally. Nuclear power is simply too expensive, and it can reduce far more CO_2 far more cheaply by simply providing power directly to the grid. It also already achieves a benefit often touted for future hydrogen projects of providing electricity that isn't intermittent.

But nuclear power plants, like most fossil fuel power plants, generate a considerable amount of heat, which they throw away. This appears to create an opportunity because high-temperature electrolysis, 900–1,800°F (500–1,000°C), using solid oxide electrolyzer cells, can, in theory, achieve very high efficiency. The DOE's Idaho National Laboratory has for years been aggressively promoting such high-temperature electrolysis as "a path to hydrogen economy."[54]

Yet the technology is not yet commercial and will require major technical advances to get there, as one November 2023 report explained.[55] Given all the other challenges that hydrogen faces in becoming economically and environmentally useful, it seems unlikely that this would be a good use of the heat and power from a nuclear power plant. Indeed, if somehow you could make an affordable small nuclear plant that you could co-locate with a major industrial facility such as hydrogen generation, why not just locate it with an industry that could use the high-temperature steam from the SMR to directly replace steam generated by natural gas? After all, that is one of the few uses for carbon-free hydrogen being seriously explored.

But there's another major limitation of high-temperature electrolysis. Safety is perhaps the single highest priority for any nuclear power plant. The NRC devotes significant time and money to vetting proposed designs for nuclear power plants or any major change in such designs. Regulators understand the risks posed by the uniquely hazardous physical characteristics of hydrogen because nuclear accidents often generate massive amounts of hydrogen.

Indeed, the 1979 President's Commission on the Three Mile Island (TMI) accident concluded, "Those managing the accident were unprepared for the significant amount of hydrogen generated during the accident."[56] During the licensing process, "the utility represented and the NRC agreed" that in a major "loss of coolant accident," hydrogen would not be a concern for weeks. But in reality, "in the first 10 hours of the TMI accident," which was not even considered a major accident, "enough hydrogen was produced in the core by a reaction between steam and the zirconium cladding and then released to containment to produce a burn or an explosion that caused pressure to increase by 28 pounds per square inch in the containment building."

In 2011, the International Atomic Energy Agency in Vienna explained in a 160-page report on the dangers of hydrogen in severe accidents that "ignition of dry hydrogen–air mixture can occur with a very small input of energy."[57] How small? "Possible sources of accidental ignition are numerous, such as sparks from electrical equipment and from the discharge of small static electric charges." That's why, in the few industrial settings where hydrogen is plentiful, safety rules are strict and can be onerous. For instance, people are often required to wear static-free clothing.

Hydrogen is unusually flammable compared with other gases. In a mixture with air, the lower flammability limit is a 4% hydrogen concentration, and the upper limit is 75%.[58] The flammability range of natural gas in air, by contrast, is only about 5%–15%. Hydrogen burns invisibly. The normal flame speed of hydrogen is 5–10 feet per second, some ten times faster than that of natural gas. All of this makes hydrogen fires harder to control.

Hydrogen has another unusual property of particular interest to the International Atomic Energy Agency. It has many combustion regimes "in a severe accident scenario." In particular, for concentrations "above

10%, acceleration up to sound velocity has been found in many experiments." The velocity of sound is about 1,000 feet per second. But scenarios with higher concentrations of hydrogen, such as 30%, can create a detonation, "a combustion wave that travels at supersonic speeds relative to the unburnt gas in front of it." That speed can be over 6,000 feet per second.

Therefore, we should expect the NRC and other regulatory bodies to apply significant oversight of any design or design change that places a large hydrogen generation and storage system near a nuclear reactor. Such systems add entirely new risks to a nuclear plant, since they can have severe explosive accidents of their own, which is why industrial settings typically require large separations between them and other buildings.

A November 2023 Clean Air Task Force study explained, "Actually harnessing nuclear process heat from an existing reactor for electrolysis is a lot more complicated than conducting a simulation."[59] As one example, "tapping into a boiling water reactor's steam system would require adding a radiation-shielded heat exchanger loop. *This could quickly become a permitting nightmare*, given the regulatory and safety requirements placed on nuclear plants, particularly in the United States."

Even in other countries, and with "pressurized water or gas-cooled reactors which have a shielded loop," the challenges in coupling a solid oxide electrolysis system with a reactor are enormous. Again, hydrogen generation systems require strict safety rules to minimize the chance that a catastrophic accident will occur or, if it does, that it will endanger nearby buildings. France's state-owned electric company, EDF, has pursued such a project and described serious limitations facing similar retrofits. For instance, the potential to harness steam is limited to lower pressures, "significantly limiting the size of the electrolyzer to double-digit MWs." Also, the electrolyzer needed to be "sited at least several hundred meters from the reactor for safety reasons, limiting the quality of heat that can be transferred to the [solid oxide electrolyzer cell] to about 200C."

Thus, large-scale hydrogen generation using nuclear plants located with high-temperature thermal electrolysis seems fraught with risk, and efforts to eliminate that risk probably undercut nuclear-generated hydrogen's modest value. Ultimately, using nuclear power plants of any size to make hydrogen for use as an energy carrier is unlikely to be a practical, affordable, or scalable strategy, and it raises serious issues of safety and opportunity cost.

CHAPTER 5

The Hype About Dirty Blue Hydrogen

FOR HYDROGEN TO SHIFT FROM BEING A CLIMATE PROBLEM to a solution, we must quickly replace "gray" hydrogen made from natural gas via a CO_2-emitting process called steam methane reforming. That's how nearly half of all global hydrogen and 95% of US hydrogen are made.

The oil and gas industry has been pushing a strategy euphemistically called "blue hydrogen," adding a carbon capture and storage (CCS) system to the reformer that will, in theory, capture a high fraction of the CO_2 emitted, then compress it and transport it via pipeline to permanent underground geologic storage.

But the chances are effectively zero that hydrogen produced this way can meaningfully contribute to the climate fight by 2050 (if ever). In contrast, the chances are high that blue hydrogen lies somewhere between a dangerous distraction and an outright boondoggle but will, either way, undermine real climate action.

There isn't "anything in the peer-reviewed literature to support" the claims that "blue hydrogen is low-emissions or zero-emissions," as one expert noted in 2023.[1] A 2023 US Department of Energy (DOE) study found blue hydrogen wasn't even as clean as hydrogen from gasified coal with CCS.[2] Another 2023 analysis concluded, "Blue hydrogen is not clean or low-carbon and never will be."[3] A key reason is the leakage of methane, a potent greenhouse gas (GHG), in the production and delivery of natural gas. So the term "blue hydrogen" is really greenwashing, and it should be called dirty blue hydrogen.

In addition to its lack of environmental benefits, there's no evidence dirty blue hydrogen will make economic sense any time soon. Green hydrogen projects can be small and modular, to fit growing demand. But for dirty blue hydrogen "a typical project costs $500 million to build," warns a 2021 Bloomberg article pointedly titled "Blue Hydrogen Could Become the White Elephant on Your Balance Sheet."[4] It could become an asset so expensive to operate and maintain that it never becomes profitable. As a result, there is "an awful lot of risk to bear on a single project. Most companies won't have the appetite for it."

The risk is especially high because there isn't much demand for clean hydrogen at that scale and probably won't be for years. Moreover, dirty blue hydrogen isn't actually clean, so why would buyers pay a premium for it? After all, green hydrogen can potentially get much cheaper, especially as renewable power continues to get cheaper. But "blue hydrogen costs have little scope to fall," especially in places such as Europe with high gas prices. "So the large-scale of blue hydrogen is actually a curse," explains Bloomberg. Indeed, a major 2022 report warned that "investments in supply chains based on fossil fuels (blue or grey)—especially assets expected to be in operation for many years—may end up stranded."[5]

The Many Intractable Problems of Dirty Blue Hydrogen

Several intractable problems must be solved before hydrogen from natural gas with CCS can ever be a real climate solution. This chapter focuses on the problems with CCS. The problems with carbon capture will also be important in understanding the challenges facing e-fuels because most of those synthetic liquid fuels made from hydrogen are also made from CO_2 via carbon capture. But the biggest overarching problem for dirty blue hydrogen is that for it to even be a contender as source of near-zero-GHG hydrogen, the nation and the world would need a major push to sharply reduce leakage from natural gas drilling and delivery, ideally by a factor of ten.

Natural gas is mostly methane, a GHG that is eighty times as powerful at trapping heat as CO_2 over a twenty-year period, which is the key timeframe to avoid crossing the dangerous temperature thresholds in the Paris Climate Agreement. A July 2023 journal study found methane leaks can make gas as dirty as coal from a GHG perspective.[6] Current leakage rates in the United States of a few percent are a fatal limitation to dirty blue hydrogen as a potential climate solution.[7]

Moreover, natural gas shipped from the United States to Europe and elsewhere in the form of liquefied natural gas (LNG) would be worse as a feedstock for dirty blue hydrogen. It has a higher global warming potential than coal because of the energy needed for liquefaction, transport, and delivery, plus the potential for more leakage during these stages, as an October 2024 study in *Energy Science & Engineering* concluded.[8] A 2019 study from the DOE's National Energy Technology Laboratory also found significant GHG emissions associated with LNG.[9] Natural gas imported from Russia to countries in Europe and Asia is problematic, too, because of the high leakage rates associated with Russian natural gas infrastructure.

So unless the nation and the world can bring leakage rates down sharply—and there has been little evidence of progress in the past decade—putting a CCS system on a methane reformer is a waste of money. The emissions from the methane leaks simply overwhelm any potential savings.

Even if the leakage problem does get solved, CCS itself still has a number of intractable problems. First, a practical and cost-effective system must be commercialized to capture 99% of the CO_2 emitted during the methane reforming process. But no such commercial system for dirty blue hydrogen exists today, nor is there any reason to believe such systems will be practical, affordable, and scalable for the foreseeable future. The few gas reformers with CCS that do exist today are small and capture at most 60% of the CO_2 emissions, as a December 2023 DOE report noted.[10] But capture rates can also be as low as 29%.[11]

The commercialization of natural gas CCS has gone even more slowly than coal CCS and is only at the stage of design and small-scale demonstration. One small 40-megawatt CCS system ran at a natural gas plant in Bellingham, Massachusetts, from 1991 to 2005. "If you had told me in 2011 that there would not be a single commercial-scale gas-fired power plant with CCS anywhere in the world," wrote Oliver Morton, *Nature*'s former chief news and features editor, in 2021, "I would either not have believed you or would have found what you said very depressing."[12]

Second, the CO_2 captured from dirty blue hydrogen systems must *not* be used for enhanced oil recovery (EOR). Over 70% of current global CCS systems capture CO_2, put it under high pressure, and transport it to existing oil (or gas) wells, where it is injected to squeeze more fossil fuels out of the ground. In a world that must reduce CO_2 emissions to as close to zero as possible by midcentury, using captured CO_2 to extract otherwise uneconomical oil is a climate problem, not a climate solution.

Third, the pipeline siting problem would have to be solved since it has become exceptionally difficult to build new pipelines, as the *New York Times* explained in a 2020 article titled "Is This the End of New Pipelines?"[13] And those battles have already spread to CO_2 pipelines, as the *Wall Street Journal* detailed in a September 2023 article, "A New Nimbyism Blocks Carbon Pipelines."[14]

Fourth, a measurement, reporting, and verification (MRV) system must be put in place to ensure that all of the captured CO_2 actually gets pumped to a certified geologic repository and to track what actually happens to that buried CO_2. That will require a quantum leap in MRV. In this country, oil companies claim the most tax credits for CCS while simultaneously fighting efforts to require MRV of CO_2 storage. As a result, "the issue of companies claiming credits for unverified tons of captured carbon is rampant in the United States," as one group of climate experts and academics has explained.[15]

Significantly, as with using renewables to make hydrogen, there is an opportunity cost to trying to build expensive and difficult-to-deploy CCS systems to make hydrogen. After more than two decades of trying, large-scale CCS systems have not been successful in the marketplace. The world's top experts on climate science and solutions concluded in their 2022 review of the scientific literature that "implementation of CCS currently faces technological, economic, institutional, ecological-environmental and socio-cultural barriers."[16] This conclusion of the UN Intergovernmental Panel on Climate Change (IPCC) was signed off on by all the world's nations.

At the same time, scaling up CCS will be immensely challenging. Total global GHG emissions are 50 billion tons of CO_2 equivalent. Yet sequestering just 3 billion tons of CO_2 a year would mean capturing, transporting, and storing a volume of compressed CO_2 greater than the more than 90 million barrels of oil a day extracted by the global oil industry, which took a century to develop. As one expert put it, "Needless to say, such a technical feat could not be accomplished within a single generation."[17]

Princeton University's net zero by 2050 pathway includes an additional 60,000 miles of pipelines dedicated to CCS. That figure isn't very plausible to start with, but we certainly can't waste time, money, and effort trying to get CO_2 pipelines sited and built for something as problematic as dirty blue hydrogen.

Suppose we actually do figure out how to make CCS systems practical, affordable, and scalable, including a large number of new pipelines.

The best use of such systems would be directly capturing, transporting, and storing current CO_2 emissions from existing industrial facilities, including power plants. That would certainly achieve far more reductions at far lower cost than trying to turn dirty blue hydrogen into a commercially successful low-carbon energy carrier. That's especially true since even if you could solve all the intractable problems discussed above, it's unclear we can solve hydrogen's many other intractable problems that make it such a lousy energy carrier.

Despite dirty blue hydrogen's many fatal limitations, the nation and the world are on track to squander tens of billions of dollars on it. For instance, of the seven US "hydrogen hubs" aimed at jumpstarting clean hydrogen production—each funded at roughly $1 billion—five of them are planning major investments in blue hydrogen, Canary Media noted in October 2023.[18] That strategy defies logic, especially given that such hubs will probably each need CO_2 pipelines—and hydrogen pipelines—and it is increasingly difficult to site a single pipeline in this country.

Who stands to win the most if we invest billions of dollars in blue hydrogen? In December 2023, Yasir Al-Rumayyan, chair of the world's largest oil company, with revenues exceeding $500 billion a year, said that Saudi Aramco will be a major investor in blue hydrogen production.[19] He is also the governor of the $900 billion Saudi Public Investment Fund. Support for blue hydrogen from the fossil fuel industry has deep pockets. As Canary Media explained, fossil fuel companies "form the core membership of the biggest hydrogen lobbying groups."[20]

The Gulf Cooperation Council (GCC), which includes Saudi Arabia, Qatar, and the United Arab Emirates, "has a head start in CCUS" according to a December 2023 report, "Carbon Capture, Utilisation and Storage in the GCC."[21] The report notes that the CCS facilities in those three countries are focused on EOR. But adding EOR to dirty blue hydrogen is abandoning any pretense that the goal is to avoid catastrophic climate change. The real goal of that strategy was summed up in the title of a 2024 Columbia University analysis, "How Carbon Capture Technology Could Maintain Gulf States' Oil Legacy."[22]

The rest of this chapter examines the challenges facing CCS more closely. This technology is relevant not just to dirty blue hydrogen but also to the e-fuel discussion in Chapter 6 and industries far beyond hydrogen production. With tens of billions of dollars of funding on the horizon—and the opportunity to appear to support climate action while continuing to produce fossil fuels—dirty blue hydrogen and CCS are the linchpins of the fossil fuel industry's plans for hydrogen and climate change as a whole.

An Introduction to CCS

CCS systems recover carbon dioxide from the emissions of industrial facilities (such as power plants and natural gas processing facilities) and sequester it underground. CCS does not remove CO_2 from the air; it only reduces the amount of CO_2 put into the air. In contrast, carbon dioxide removal reduces the absolute amount of CO_2 in the atmosphere (see Chapter 6).

We've had commercial CO_2 capture since the 1970s, deployed primarily in gas processing plants to separate CO_2 for use in EOR.[23] Natural gas, as it comes out of the ground, can have much higher levels of CO_2 than are permitted for pipeline gas. So processing plants are used to remove the CO_2 (and other contaminants) in raw natural gas. Those separation systems account for most (about 70%) of the CO_2 captured each year worldwide.

The full CCS process has three parts:

1. Separating CO_2 from all of a plant's other emissions to create a highly concentrated stream,
2. Compressing the CO_2 to more than 1,000 pounds per square inch to transport it, typically via pipeline, to the storage site, and

3. Injecting that CO_2 into an underground geological repository (and monitoring and verifying the integrity of the storage system).

Not one of those three parts has been demonstrated to be practical, affordable, scalable, and sustainable.

To run a CCS system, power plants and other industrial facilities generally must add an expensive postcombustion carbon capture, compression, transportation, and storage system that requires more power to run than power plants without CCS. This extra power is called the *energy penalty* or *parasitic load*. It is estimated to be 20% to 30% of a power plant's capacity.[24]

The energy penalty must be paid by either burning more fuel or cutting the net amount of power supplied by the plant. An analysis by Harvard and Massachusetts Institute of Technology researchers found that retrofitting all US coal-fired plants would increase annual coal use by 400 to 600 million metric tons or cut their net power output by 75–100 gigawatts.[25] That's as much as twice California's peak electrical demand.

Putting a carbon capture system on a methane reformer means you will have to buy and burn more natural gas to make dirty blue hydrogen than you did to make the regular dirty gray hydrogen. But that means more methane leakage, which counteracts much of the benefit of the CCS systems, especially if it doesn't capture 95% or more of the CO_2. There is another device for generating hydrogen from natural gas called an auto-thermal reformer from which it might be easier to capture CO_2, but that reformer uses much more electricity, so unless you run it on renewables that are new, local, and hourly matched (as discussed in Chapter 3), it will be dirtier still. I will focus primarily on steam reforming in this book because studies suggest it is likely to be cheaper and cleaner.

Capturing the CO_2 is often the largest cost component in the CCS system, "accounting for as much as 75% of the project cost," as the National Petroleum Council explained in a major 2020 CCS report.[26]

Sometimes, instead of *CCS* the term *CCUS* is used, which stands for "carbon capture, utilization, and storage." But a 2017 study in *Nature Climate Change* concluded, "It is highly improbable [CCUS] will account for more than 1% of the mitigation challenge, and even a scaled-up enhanced oil recovery (EOR)-CCS industry will likely only account for 4–8%."[27] So although CCS EOR "may be an important economic incentive" for some early CCS projects, "ultimately CCUS may prove to be a costly distraction, financially and politically, from the real task of mitigation." Some groups use *CCS* and *CCUS* interchangeably. Here I will generally use the term *CCS*, even when the CO_2 is used for EOR.

CCS Systems Keep Missing CO_2 Capture Goals

"So far, the history of CCUS has largely been one of unmet expectations. Progress has been slow and deployment relatively flat for years,"[28] explained the International Energy Agency (IEA) in November 2023. "This lack of progress has led to progressive downward revisions in the role of CCUS in climate mitigation scenarios."

Before being shut down in 2017, the Kemper County coal plant, "clean coal's flagship project,"[29] was three years behind schedule and billions over budget. Equipment to gasify coal and capture two thirds of the CO_2 "never worked as designed." The plant ran on natural gas for most of the prior three years and ultimately captured "only a tiny amount of carbon dioxide."

The Petra Nova CCS system was designed to capture only 40% of the CO_2 emitted from one of the many boilers of the Parish Power station in Texas.[30] Before it shut down what was then the world's largest

CCS facility in 2020, it routinely failed to come close to its annual CO_2 targets primarily "because it ran for far fewer days than expected due to myriad system challenges."[31] Also, the Petra Nova team built a natural gas plant to provide 45 megawatts to run the CCS system. When you factor in the CO_2 emitted to power the CCS system, the net capture rate was "6.2% of the Parish plant's 2015 emissions."[32] And the icing on this cake is that the captured CO_2 was used to boost production at a nearby oil field by up to 5 million barrels of oil a year.

Consider the Boundary Dam 3 coal plant in Saskatchewan, Canada. In its first three years, "the carbon capture facility had only achieved its design capacity for three days," according to a 2021 report.[33] In 2022, *E&E News* reported, "The world's sole carbon capture project on a large power plant caught 43% fewer metric tons of carbon dioxide in 2021 compared with the year before."[34] The plant was supposed to capture 90% of the CO_2 it produces, but SaskPower has since reduced that goal to 65%.

In 2021, Chevron said its $54 billion Gorgon LNG facility in Australia would fall far short of its promised CCS goal.[35] The company had committed to capturing 80% of its emissions over a five-year period starting in July 2016 as "a condition of the development's state approval." But it appears Chevron captured well under half of what was promised.

When looked at in the aggregate, carbon capture projects have performed similarly poorly. The Climate Council, which consists of some of Australia's leading climate scientists and policy experts, pointed out in 2021, "There is not a single carbon capture and storage project in the world that has . . . captured the agreed amount of carbon."[36] In 2022, a study of storage rates of twenty large CCS projects since 1996 compared "widely reported capture capacity" figures with estimates of the amount of CO_2 that was actually captured and stored geologically (based on public data).[37] Its conclusion: "The widely used capture capacity is in aggregate 19–30% higher than storage rates and is not a good proxy for estimating storage volumes."

Carbon Storage

Once CO_2 has been captured and compressed, it must be stored in a permanent location. By far the most used geologic storage sites today are existing oil (and gas) wells for EOR. There are now more than 100,000 EOR wells injecting CO_2 in the United States.

But the geologic sites that contain the largest potential capacity to store CO_2 are deep saline repositories and sedimentary basins a mile or more underground containing unpotable brines—brackish water unsuitable for agriculture or drinking. Deep saline reservoirs can be found onshore and offshore, much as oil and gas wells can. These reservoirs are estimated to have the capacity to store more than 8,000 gigatons of CO_2 in the United States alone, which is some forty times the capacity of oil and gas wells. Yet there are many complicating factors at work, which means that we don't actually know how much CO_2 could be stored permanently, safely, and verifiably in these reservoirs. It could be much less.

One such factor is the geology of the oil and gas wells in comparison to the deep saline reservoirs. Because we already know that oil fields can trap hydrocarbons, we have "greater confidence in the trapping mechanism at CO_2-EOR projects," as one analysis put it.[38] At the same time, because of "the long operating histories and large number of wells in the oil field setting," the geology of EOR sites is probably much better understood than that of saline aquifers. Also, it is easier to keep the pressure relatively constant at an EOR site since as the CO_2 goes in, oil comes out. At saline sites, the pressure is much more likely to rise, unless some way is found to remove the saline fluids to balance things. Ultimately, it is not clear that our experience with EOR is helpful for understanding non-EOR geologic storage.

The Problems with CCS for EOR

Right now, there is very little dirty blue hydrogen, in part because it is so uneconomical. That's precisely why Saudi Arabia, Qatar, and the United Arab Emirates are pursuing dirty blue hydrogen with EOR, because that extra income from using the captured CO_2 to produce more oil helps make the whole system more affordable.

According to the IEA, EOR can as much as double the amount of petroleum that can be recovered from an oil field.[39] From a climate perspective, EOR is problematic for several important reasons.

First and foremost, using a ton of CO_2 for EOR is much less beneficial to the climate than sequestering a ton of CO_2 in a deep saline aquifer, partly because of the emissions from the extra oil recovered. A 2015 analysis by the IEA found that for every ton of CO_2 delivered to an EOR field, only 63% is a net CO_2 emission reduction.[40]

In fact, EOR may not lead to a reduction in CO_2 emissions. A 2020 review of over 200 studies makes a compelling case that a truly comprehensive lifecycle analysis shows EOR to be a net *increase* in atmospheric CO_2, not a decrease.[41] One commonly cited study of coal plants with CCS used for EOR found that "between 3.7 and 4.7 metric tons of CO_2 are emitted for every metric ton of CO_2 injected" underground.[42]

The second problem with EOR is that in the United States, a ton of CO_2 for EOR is currently treated the same as a ton stored permanently without generating more oil. That means a company not only can make money selling the CO_2 to the EOR industry but also can get paid for the (supposed) reduction in CO_2 emissions, through a tax incentive or the rapidly growing global carbon market.

What does this mean in practice? In 2008, the US Congress launched the corporate income tax credit for CCS. Congress has repeatedly expanded and increased it. In 2019, the Clean Air Task Force modeled the impact of the tax credit on CCS and found that "the additional revenue from EOR creates more favorable economics for CCS deployment relative to storing CO_2 in saline reservoirs. In the real world, these results, would mean rapid EOR infrastructure development."[43] In fact, "the tax credits did not result in any saline storage." The EOR crowded out the competition for other types of CO_2 storage.

The Inflation Reduction Act of 2022 made EOR even more profitable, roughly doubling the tax credit. It raised the tax credit for CCUS by power plants and industrial facilities to $85 per metric ton if the CO_2 is stored in geologic formations and $60 if the CO_2 is either used beneficially (e.g., to make a commercial product) or for EOR.

A lot of oil can be recovered through CO_2 injection. One 2018 market analysis estimated that the total amount of extra oil accessible by EOR in the United States alone is 284 billion barrels.[44] That's almost a forty-year supply of oil at the current level of US consumption, and some estimates are even bigger.[45] This does not represent a move away from building out more fossil fuel infrastructure, something that both the IEA and IPCC say is crucial if we are to meet the Paris climate targets.

In fact, a May 2022 study, "Existing Fossil Fuel Extraction Would Warm the World Beyond 1.5°C," found that staying within a 1.5°C carbon budget "implies leaving almost 40% of 'developed reserves' of fossil fuels unextracted."[46] Developed reserves are "the portion of fossil fuel reserves within actively producing or under-construction oil and gas fields and coal mines." So we need to shut down a substantial portion of existing oil fields, not subsidize their further development.

The third problem with EOR is that it can store only a small fraction of global emissions. Deep saline aquifers can potentially store forty times as much CO_2 as EOR. So again, the overwhelming majority of CCS systems operational today aren't helping us learn how to scale up carbon storage, and giving EOR an economic advantage over saline aquifers will only slow the needed transition.

A detailed discussion of EOR by *Vox* in 2019 noted another problem with EOR.[47] Fossil fuel corporations are huge companies driven to produce and sell as much hydrocarbon as possible, a major reason they have had a poor track record on climate for decades. So anyone "who believes oil and gas companies will be good-faith partners in a climate-emergency effort, is indulging in a kind of willful naivete." And oil companies do more CCS than any others—and Exxon does the most of all.

In the specific case of EOR, oil and gas companies have already been bad-faith actors. Exxon "has relentlessly fought EPA [Environmental Protection Agency] oversight" of EOR, *Inside Climate News* reported in 2020.[48] Through the industry-funded Energy Advance Center, *Vox* noted, the companies "submitted comments to the IRS arguing that the agency should get rid of the strict verification rules 45Q requires for EOR sequestration." The companies want to be able to claim those credits strictly on the basis of how much CO_2 was *delivered* to the EOR site, without any MRV of how much was actually stored.

It turns out many of these companies have already been claiming credits without any of the required MRV, as a 2020 investigation by the Inspector General for Tax Administration at the US Department of Treasury uncovered.[49] The investigation found that 87% of the total 45Q credits claimed—nearly $900 million in credits—did not comply with the rules requiring companies to "develop and implement an EPA-approved site-specific MRV plan."

In 2022, a group of over 400 "scientists, academics, and energy system modellers" signed a letter strongly opposing the creation of a Canadian tax credit for CCS.[50] As the group describes it, "The issue of companies claiming credits for unverified tons of captured carbon is rampant in the United States under Section 45Q." The oil and gas industry is acting in bad faith—and trying as hard as possible to kill even the most basic MRV efforts, as we'll see.

In March 2024, Reuters reported that Summit Carbon Solutions, which is trying to build the country's biggest CO_2 capture pipeline, has pledged many times it won't be used for EOR.[51] As of August 2024, the company's website still states, "The Summit Carbon Solutions project will not be used for enhanced oil recovery."[52] EOR is not a politically popular way to use such a pipeline, so saying otherwise could cost it crucial public support. But Summit has given North Dakota's oil sector a different message, "according to a Reuters review of state regulatory filings and recordings of public appearances by company executives." The article notes the oil industry wants the pipeline for EOR "to reverse the once-booming region's flagging output." Reuters then quotes Summit's executive vice president at a December 2023 event in Bismarck: "Today, we don't have any shippers who want to ship CO_2 for EOR. When that changes, we will likely move it for that purpose."

The Problems with Pipelines

Even if CO_2 is captured at high enough rates to make financial and environmental sense, the compressed carbon must be transported before it can be stored for good. JP Morgan noted in its *2021 Annual Energy Paper*, "just to sequester an amount equal to 15% of current US GHG emissions, would require infrastructure whose throughput volume would be higher than the volume of oil flowing through US distribution and refining pipelines, a system which has taken over 100 years to build."[53]

Yet compared with the last 100 years, we are now in a political climate where it is increasingly difficult to build even a single new pipeline.[54] In October 2023, a developer of a planned 1,300-mile CO_2 pipeline network winding through the farm belt canceled the project.[55] The company cited the "unpredictable nature of the regulatory and government processes" in two out of the five states the pipeline had to cross.

A 2020 Princeton study of net-zero-by-2050 scenarios found that just to capture and store nearly 1 gigaton of CO_2 a year, the United States alone might need to build over 60,000 miles of new pipelines.[56] Is this really a scalable option, especially for a technology that is not even commercial today?

Public concerns about CO_2 pipelines are pervasive. A 2019 review article of 135 studies on public perception of CCS found that the most mentioned risk was "CO_2 leakage, migration (from storage or pipeline) or explosion."[57] The public's concerns are justified: The IPCC pointed out in 2021 that "there is a non-negligible risk of carbon dioxide leakage from geological storage and the carbon dioxide transport infrastructure."[58] Neither these risks—nor the public's perception of them—can be addressed without much more stringent MRV than many if not most CCS facilities currently employ, as discussed below.

Dangerous leaks happen in part because of how CO_2 is transported and stored. CCS requires a vast amount of CO_2 to be injected deep underground under high pressure, above 1,000 pounds per square inch. This is sufficient to turn CO_2 into dense phase or supercritical CO_2, with a density similar to a liquid (at least half the density of water) but a viscosity (resistance to flow) similar to a gas. The supercritical phase is generally used for both the transportation of CO_2 and injection because it minimizes the overall volume of the CO_2.

Because of these properties, carbon dioxide pipelines have to be rated for much higher pressures than natural gas pipelines. Compared with gas compressors, "the chemical and physical properties of CO_2 require modifications to compressor design, construction materials, and sizing," notes a 2019 National Petroleum Council Report.[59] For this reason, the council concluded that "it is not anticipated that repurposing existing natural gas pipelines would significantly help develop" an expanded US pipeline network for CO_2.

At the surface, a supercritical CO_2 leak can create a huge gaseous cloud that "can sometimes hang in the air for hours."[60] It is an asphyxiant and intoxicant—and odorless and colorless. So it can be more dangerous than chemicals that may rapidly disperse into the air.

Problems do occur with CO_2 pipelines, especially since CO_2 is so highly corrosive if contaminated with even a little water that it will eat through the carbon steel used in oil pipelines. According to data from the Pipeline and Hazardous Materials Safety Administration (PHMSA), which is part of the US Department of Transportation, there have been sixty-six incidents on CO_2 pipelines, with no fatalities.[61] An Association of Oil Pipe Lines analysis found that there were "1.1 CO_2 pipeline incidents per 1,000 miles." That is twice the incident rate from 1990 to 2002.

Pipelines carrying supercritical CO_2 are "particularly susceptible" to dangerous "running ductile fractures," according to a 2022 report for the Pipeline Safety Trust.[62] These are "particularly dangerous fractures that can 'unzip' a CO_2 transmission pipeline for extended distances exposing great lengths of the buried pipeline." They generate "extreme rupture forces" that "throw tons of pipe, pipe shrapnel, and ground covering, generating large craters along the failed pipeline."

A major pipeline incident shows what can happen when a CO_2 pipe bursts. In February 2020, a 24-inch high-pressure pipeline containing CO_2 and some hydrogen sulfide burst in Satartia, Mississippi, as *HuffPost* reported.[63] As a result, many people felt dizzy, nauseous, and disoriented. Nearly fifty people were hospitalized, and over 300 were evacuated.

"We got lucky," the Yazoo County Emergency Management Agency director who oversaw the response effort told *HuffPost*. "If the wind blew the other way, if it'd been later when people were sleeping, we would have had deaths." A nine-month *HuffPost*/Climate Investigations Center investigation of this incident revealed a number of concerning aspects. For instance, the pipeline owner, Denbury, had another mass leakage on October 7, 2020, while reconnecting the damaged part of the pipeline and attempting to remove any remaining CO_2 from the section, releasing even more carbon dioxide.

The PHMSA has been slow to release an incident report on the cause of the rupture. "That incident happened over two years ago," Bill Caram, executive director of the Pipeline Safety Trust, told *Grist* in 2022.[64] "It's crazy that communities are being asked to bear the burden of the risk of these pipelines when this report sits unreleased with all these unanswered questions."

This suggests PHMSA is not doing an adequate amount of oversight of CO_2 pipelines, a point underscored by the 2022 Pipeline Safety Trust report. That report asserts current CO_2 pipeline regulations "are incomplete or in conflict with" the massive scale-up in CO_2 for CCS. "As a result, many of PHMSA's regulations no longer are adequate to protect public safety." The report offers a number of suggested changes and concludes, "The country is ill prepared for the increase of pipeline mileage being driven by federal CCS policy."

Indeed, thousands of more miles of CO_2 pipelines are in the works, including three multi-billion-dollar projects in the Midwest, thanks to the repeated boosts in the tax credit for CCS and a program of low-interest loans for such pipelines that passed into law in late 2021. Making a real dent in US CO_2 emissions would require scaling up more than tenfold from the current 5,000 miles of CO_2 pipelines in the country. At the current incident rate, such a network would have over seventy pipeline incidents a year.

Yet as the IPCC's 2005 report on CCS notes, "If CO_2 is transported for significant distances in densely populated regions," then "public concerns about CO_2 transportation may form a significant barrier to large-scale use of CCS."[65]

The Storage Challenge

Injecting vast amounts of high-pressure CO_2 into deep underground geologic formations has its own unique challenges. In addition to the health threats directly to humans from CO_2 leaks is the threat to groundwater.

A 2012 study explains that after injection, "supercritical CO_2 remains highly mobile, buoyantly flowing upward and significantly extending outwards along the sealing layer."[66] The sealing layer or caprock is an impermeable layer of rock above the more porous rock the CO_2 is stored in under high pressure. The buoyancy of CO_2 creates "much of the security risk" in geologic storage, notes a 2015 study,[67] "providing a driving force for it to escape back to the surface via fractures, or abandoned wells."

When CO_2 enters an aquifer, some of it dissolves into the groundwater, creating carbonic acid. This acidification "can then cause an increase in the mobilization of major and minor elements as well as potential contaminants," explains a 2015 review by the DOE's Pacific Northwest National Laboratory (PNNL).[68]

The PNNL review examined fourteen studies in the previous decade assessing how potential CO_2 leakage from deep storage reservoirs could affect the quality of overlying freshwater aquifers. Eight of the studies indicate "CO_2 leaks pose a serious risk," four found low levels of risk, and two found "some possible benefits." PNNL then did its own field experiments and worked with Lawrence Berkeley National Laboratory on modeling efforts. The labs found "there are cases where there may be measurable, long-term impacts to groundwater due to CO_2 or brine leaks."

What is the risk to the public? Injected CO_2 could bubble up into near-surface drinking water aquifers, "driving up levels of contaminants in the water tenfold or more," according to a 2010 Duke University study.[69] The researchers warned, "Potentially dangerous uranium and barium increased throughout the entire experiment in some samples." The study determined geologic criteria that need to be monitored or avoided entirely.

But monitoring by itself may not satisfy many local residents. The fact that you may be able to detect a leak early isn't the issue. The issue is whether leak detection leads to actions that would stop the leak before it taints the groundwater.

As for government oversight, a 2016 Government Accountability Office report on the EPA's monitoring of injection wells found, "EPA has not consistently conducted oversight activities necessary to assess whether state and EPA-managed programs are protecting underground sources of drinking water."[70]

The groundwater contamination problem may well limit the places where sequestration is practical, either because the site's geology is problematic or because it is simply too close to the water supply.

Leaks that directly affect human health also happen. The Salt Creek Field in Wyoming, an EOR project, has seen repeated seepage to the surface of CO_2.[71] One leak shut down a local school for months after health officials detected dangerous levels of noxious gases inside. Wyoming Public Radio explained that CO_2 levels "were 26 times the recommended limit, which made some areas of the school oxygen-deficient." Levels of benzene, a very hazardous chemical, "were 200 times the amount deemed safe."[72] Eventually, the cause was determined: "There was a leaky abandoned well, sitting squarely in the middle of the schoolyard."

Blowouts and well failure may be "the largest risk associated with CO_2 injection for sequestration in deep brine reservoirs," a 2009 study on EOR risk assessment explained.[73] Two companies had a combined twelve blowouts in the past five years of operation, and a third company had twelve blowouts in the past ten years. A well blowout happens when the well operator loses control of the pressures in the injection system, causing fluid to flow out, often rapidly. A 2003 article in *World Oil* pointed out that recently "there has been an increased frequency of CO_2 blowouts in injection projects."[74] A 2014 study of three Texas districts cataloged 158 blowouts from 1998 to 2011.[75]

In 2012, ProPublica examined well records and government summaries of more than 220,000 well inspections and found that "structural failures inside injection wells are routine."[76] From late 2007 to late 2010, ProPublica documented that "one well integrity violation was issued for every six deep injection wells examined—more than 17,000 violations nationally." Of those, over 15,000 violations were injection wells for enhanced recovery.[77]

Denbury, the pipeline owner in the Satartia, Mississippi, incident, has had several blowouts. In August 2011, their Tinsley Field itself suffered a big CO_2 blowout in Yazoo County that "that took 37 days to bring under control, sickened one worker, and killed deer, birds, fish and other animals."[78] A June 2013 underground CO_2 blowout at Denbury's Delhi field in Louisiana lasted over six weeks: "Concentrations of carbon dioxide were so high initially that . . . responders wore breathing apparatus to keep from suffocating."[79]

Multiple CCS projects have already been canceled because of public concern over the risk of leaks. A 2021 study concluded that "Societal acceptance has proved to be a major barrier for CCS, both in the Netherlands and elsewhere."[80] Public concern about CO_2 leaks (small and large) has impeded or killed a number of CCS projects around the world.

Concern about leaks helped kill one of the world's first full CCS demonstrations by Swedish company Vattenfall in northern Germany. The project was supposed to start burying CO_2 in September 2008. In 2009, Germany tried to introduce legislation that would have had the government assume liability for companies injecting carbon dioxide underground. It failed to pass. Vattenfall did not get a permit to bury the carbon dioxide. As a result, in July 2009 the plan "ended with CO_2

being pumped directly into the atmosphere, following local opposition at it being stored underground."[81]

Similarly, a Netherlands project to capture CO_2 from Shell's Pernis oil refinery and bury it under the town of Barendrecht in a depleted gas field was canceled in 2010 as residents "worried about damage to their homes and subsequent drops in property value."[82]

Public opposition will probably be as big a threat to the scalability of storage sites as it will be to the scalability of pipelines. By one estimate, up to sixty large storage sites will need to be discovered and developed per year for the next twenty years to meet multi-billion-ton annual storage goals.[83] In thinking of the challenge of siting hundreds of underground storage sites, remember that this country still does not have a single long-term storage facility for high-level radioactive waste—after four decades of trying—due largely to public opposition.

The Fight for MRV

Figuring out whether and how much a given geologic storage site for CO_2 might leak in one, ten, or one hundred years is not simple. The IPCC states in its 2005 *Special Report on Carbon Dioxide Capture and Storage* that for large-scale CO_2 storage projects, "the balance of available evidence suggests" the chances of retaining over 99% of the CO_2 for at least one hundred years is greater than 90%, "assuming that sites are well selected, designed, operated and appropriately monitored."[84]

The EPA regulates Class VI carbon injection wells under the Safe Drinking Water Act. Their rules to protect underground drinking water sources include extensive site characterization requirements, injection well construction and operation requirements, and comprehensive monitoring of well integrity, CO_2 injection, and storage.[85] Groundwater quality, reporting, and recordkeeping requirements are also vital.

Yet as we have seen, the oil industry is fighting to weaken all reporting and oversight of geologic storage. And many states themselves are working to strip all corporate liability from CCS—and to strip the EPA of its legal oversight role. In May 2022, Texas approved a measure to gain sole regulatory authority over the state's onshore and offshore CO_2 injection wells. North Dakota and Wyoming gained primacy under the Trump administration.

When regulation is left up to individual states, the results can be problematic. *S&P Global* reported in May 2022 that Commission Shift, a Texas nonprofit, is critical of the state "for its lax oversight and willingness to accept large campaign contributions from the companies it regulates."[86] The group's executive director said state regulatory "commissioners are on the record either deriding federal climate standards or saying they don't believe in human-induced climate change, disqualifying them from overseeing such a complex and significant technology." It seems unlikely the state's requirements for MRV of injection wells and storage would be at all comparable to EPA requirements.

The liability issue was summed up in an April 2022 headline by *Inside Climate News*: "Proponents Say Storing Captured Carbon Underground Is Safe, But States Are Transferring Long-Term Liability for Such Projects to the Public."[87] Companies such as ExxonMobil, BP, and Denbury have been "seeking protections from long-term liability." At least seven states have passed laws allowing companies to transfer responsibility for storage projects to state governments after the injection operations end. "Statutes that relieve operators of liability without due regard to existing legal principles create an incentive for sloppy management, leaks and public opposition," explained Scott Anderson, senior director of energy transition at the Environmental Defense Fund (EDF). Internal Exxon documents obtained by *Inside Climate News* reveal the company supports a regulatory "mechanism to de-risk the long-term liability obligations for stored CO_2," for a proposed Texas CCS project. An Exxon executive told a meeting of state oil and gas regulators in 2021, "We have to deal with liability once you put all this CO_2 in the ground." He said, "Ultimately the big challenge is liability."

In its 2022 comments on White House draft CCS guidance, EDF urged states against creating a "moral hazard" with liability laws "creating a situation where operators lack sufficient incentive to decrease their exposure risk because they will not face significant consequences if projects eventually fail or have negative effects."[88] EDF explains that some state liability laws will "encourage mediocre operations or worse."

Governments need to incentivize stronger MRV, not disincentivize it. If this problem isn't solved, we can have no confidence the dirty blue hydrogen or climate strategy built around CCS is actually accomplishing anything.

That's clear from the original big CO_2 storage site, Norway's Sleipner gas field in the North Sea, which has injected nearly 1 million tons of CO_2 yearly since 1996. The CO_2 is separated and captured at an offshore platform before it's injected into the 10,000-square-mile Utsira formation, a layer of porous sandstone half a mile under the sea floor that may be able to hold 600 billion metric tons of CO_2. In early 2013, Statoil (now Equinor) asserted, "The CO_2 is contained under an eight hundred metre thick layer of gas-tight cap rock and cannot seep into the atmosphere."[89]

But as the journal *Nature* reported in December 2013, while "the company has rich data about the Utsira formation, less is known about the overlying sedimentary layers."[90] That problem was addressed with data from the autonomous HUGIN submersible and other tools used by the European Commission's $14 million ECO2 research effort. The results were unexpected. At a site 15 miles north of the injection well, HUGIN found a 2-mile-long fracture running 500–650 feet deep and

30 feet wide in places. Since the disposed CO_2 had spread in a plume only 1.8 miles north from 1996 to 2013, ECO2 researchers asserted that the CO_2 "is unlikely to reach [the fracture] any time soon."

But researchers also found smaller vertical features they labeled chimneys and pipes running through the sedimentary layers. One is only 3 miles from the spreading CO_2 plume, and it goes all the way down into the Utsira reservoir. There are apparently many such previously unknown leakage paths, especially since the researchers examined only a few underwater sites. Thus, the Utsira reservoir may not be so "gas tight." One marine geologist who worked on the ECO2 project said, "We might have to appreciate that there is a much greater chance for some CO_2 to leak out."

The biggest implication of these findings may be that any company planning to store CO_2 beneath the sea will need to spend much more money on better seabed screening and MRV.[91] Costs may rise as "scientists recommend companies should use more advanced technology to monitor the structure of the seabed where carbon would be stored." Klaus Wallmann, ECO2 project coordinator, said, "The costs would be in the region of several million euros."

The kind of comprehensive modeling needed to ensure the ongoing integrity of underground CO_2 storage may be seen from the In Salah CO_2 storage project at a major gas-producing field in Algeria. One 2012 study noted that this project "is an excellent analogue to the saline aquifers that might be used for large scale" geologic storage around the world including the US Midwest, northwestern Europe, and China.[92]

The project was run by Equinor (then Statoil) and its partners BP and Sonatrach, the largest African oil and gas company. From 2004 to 2011, three injection wells buried some 3.8 million tonnes of CO_2. The project had to be shut down in 2011 because the CO_2 being pumped underground was apparently starting to fracture the rock that was supposed to cap it below ground.[93]

A 2013 study by experts at Statoil, BP, and Sonatrach revealed that ongoing risk assessments were valuable in identifying the danger of CO_2 and the loss of well integrity.[94] That study noted "the integration of the new seismic satellite data and dynamic/geomechanical models" in a 2010 risk assessment revealed the serious possibility of vertical leakage. That led to a decision in June 2010 to lower the CO_2 injection pressures. After further study injection was suspended entirely in June 2011, the decision was ultimately made to not to restart injection.

All this happened after just 3.8 million tonnes of CO_2 were injected. The IPCC and IEA are considering emission reduction scenarios in which 3,000 times that amount of CO_2 needs to be injected—or more. The monitoring needed to make these decisions included a "time-lapse seismic, micro-seismic, wellhead sampling using CO_2 gas tracers, down-hole logging and core analysis, surface gas monitoring, ground-water aquifer monitoring" and satellite data.[95]

In short, it takes a lot of ongoing monitoring, modeling, verification, and reporting to maintain the security of a geologic storage site. It is far from clear whether fossil fuel companies are prepared to do what is needed—and whether many states are either. This greatly burdens the EPA, at least in states where it maintains oversight. It suggests that the federal government should strive to keep oversight and that freeing the companies involved from liability is unwise. Liability is one of the few levers to motivate companies to put serious effort into minimizing the long-term risk of leakage or rupture.

What about other countries? Suppose Putin's Russia said it was permanently storing 100 million tons of CO_2 a year underground and wanted other countries to pay it billions of dollars. Would anyone trust their monitoring and reporting? Probably not. The potential for fraud and bribery is simply too enormous. But would anyone trust China or Saudi Arabia? Or ExxonMobil?

We need to set up an international regime for certifying, monitoring, verifying, and inspecting geologic repositories of carbon, like the UN weapons inspections program. There must be very high confidence worldwide that a company or country is injecting the amount of CO_2 it claims to and that there are no major leaks and no indications of a threat to storage integrity. Absent that confidence, few will be willing to pay for CCS. The problem is that, in this country at least, we appear to be moving away from regulatory oversight of CCS.

CHAPTER 6

The Hype About E-Fuels and Direct Air Capture

"THERE IS SO MUCH OVERHYPE AROUND AIR CAPTURE," Dr. David Keith told me in 2022. Keith, who had founded Carbon Engineering, a top direct air capture (DAC) company, and now leads the University of Chicago's Climate Systems Engineering Initiative, added, "The public discussion is disconnected from reality."[1]

As we've seen, the hydrogen discussion has followed the same path. An important but polluting chemical feedstock and inefficient energy carrier has been touted as a miracle solution in a great many economic sectors that it makes little sense for, such as vehicles or heating. But it has become increasingly clear that hydrogen from carbon-free electricity can't compete with simply using that same electricity to reduce emissions directly.

So hydrogen has become the classic problem in search of a solution—in search of some economic subsector where it can reduce greenhouse gas (GHG) emissions but electricity cannot. The focus of this chapter is the most dubious of all such proposed solutions, electro-fuels or e-fuels: synthetic liquid fuels that could, in theory, substitute directly for fossil fuels in ships and airplanes. All e-fuels are made using hydrogen, which itself must be made from carbon-free electricity to be a scalable climate solution, as explained in earlier chapters.

All e-fuels except e-ammonia are hydrocarbons that get their carbon either through capturing CO_2 from the air mechanically—DAC—or through capturing biogenic CO_2 from biomass. The latter typically involves putting a carbon capture system on bioenergy plants, such as biopower facilities or biorefineries that make ethanol from corn. But as we will see, DAC, much like hydrogen, has a huge opportunity cost and is not a scalable climate solution, and trying to scale up biogenic CO_2 capture would probably speed up global warming for decades.

The Bipartisan Policy Center's 2024 e-fuel primer starts its "benefits and advantages" list with "Drop-in fuel option for difficult-to-electrify sectors."[2] Certainly, long-distance aviation and intercontinental shipping are hard to electrify, although together they account for only a few percent of global GHG emissions, so decarbonizing other larger sectors is a much higher priority right now. But the discussion of e-fuels is so overhyped that countries such as Germany, along with venture capitalists, numerous startups, Porsche, and other companies, are proposing to use e-fuels as a drop-in fuel option for a much bigger sector that is already proving very straightforward to electrify: cars.

To accomplish this, they will need to integrate three complex, inefficient, and expensive processes—hydrogen production from electrolysis, DAC of CO_2, and chemical synthesis of those two gases into a liquid hydrocarbon—just to make an e-fuel that will be burned inefficiently in an internal combustion engine (ICE), generating air pollution and the original CO_2.

If that weren't challenging enough, this mammoth capital- and energy-intensive megafactory must run on renewables that are new, local, and purchased on an hourly basis to match their power consumption (see Chapter 3); otherwise, it won't be producing a carbon-neutral fuel.

This megafactory is not a simple, modular system like a solar panel, so it is unlikely to see large, steady price drops over time. Indeed, we've already seen big price hikes in new large electrolyzer projects for green hydrogen.

Yet Germany and other countries have used the idea of green e-fuels—currently "100 times more expensive than petrol," as one 2023 article noted[3]—as an excuse to oppose a long-negotiated effort to ban ICE cars in Europe by 2035.

We've already seen that, by itself, trying to run cars even on relatively cheap, dirty hydrogen is a commercial failure. But expensive green hydrogen has a massive opportunity cost because it wastes vast amounts of renewable energy that could cut far more emissions for far less money by directly replacing fossil fuels. And that's with a high-efficiency fuel cell car that emits no air pollutants.

But DAC, pulling CO_2 of out the air mechanically, must also be powered by renewable energy to be net zero emissions, and it has the same flaws. The overall efficiency of the DAC process is low (5%–10%), the price is high, and so there is also a huge opportunity cost. Those same renewables could cut far more CO_2 emissions for far less money than trying to find and extract a needle in a haystack, CO_2, whose atmospheric concentration is 420 parts per million.

In effect, e-fuels are an attempt to "unburn petrol," as Michael Liebreich, a hydrogen expert and former chair of BloombergNEF, told the *Guardian*.[4] When you burn a hydrocarbon such as oil, you add oxygen, turning the hydrogen into water (H_2O), releasing heat, and turning the carbon into carbon dioxide, which also releases heat. Highly flammable hydrogen becomes highly stable water. A dense carbon-rich liquid becomes a highly diffuse mixture of CO_2 in the air. Reversing each of those processes takes a lot of energy. As Liebreich says, "You need an insane amount of solar to do that."

If the process of making e-fuels weren't inefficient already, next you have to burn it in an inefficient engine that emits dangerous air pollutants such as NOx. A 2021 *Nature Climate Change* analysis concluded that "E-fuel mitigation costs are €800–1,200 per tCO_2," which at the time was about \$950–\$1,400 a ton.[5] All of the climate solutions funded in the Inflation Reduction Act—other than those involving hydrogen—cost less than one tenth as much per ton.

A May 2024 BloombergNEF analysis found that e-kerosene is five to nine times as costly as normal kerosene-type jet fuel. The authors offered a stunning projection: "By 2050, e-kerosene is likely to still come at a premium compared to conventional jet fuel combined with direct air capture to offset emissions."[6] In other words, it is likely to make more economic sense in a quarter century to simply keep burning fossil fuel–based jet fuel and pull a comparable amount of CO_2 out of the air with DAC than to use DAC to go the e-fuel route. Of course, by 2050 there will probably be cheaper ways to remove CO_2 from the air at scale and better sustainable low-CO_2 aviation fuels available. We don't yet know the winning strategy for decarbonizing international flights in 2050. We only know that e-fuels are unlikely to be the answer, and we should do far more research and development before making any major deployment investments.

The inefficiency and implausibility of e-fuels such as e-kerosene were captured in a calculation by the Clean Air Task Force in their November 2023 report, "Hydrogen for Decarbonization: A Realistic Assessment."[7] The word *realistic* is rarely seen in analyses of hydrogen. As the report notes, "Optimism about hydrogen's potential as a tool for decarbonization continues to drive techno-economic assessments that lack grounding in facts and experience." The report considers a simple scenario where all twenty-five daily roundtrip flights between London's Heathrow Airport and New York City's JFK are powered by e-kerosene and points out that "a facility the size of the NEOM Green Hydrogen Complex would be required just to supply the quantities of hydrogen needed to produce these fuels." That \$8.4 billion Saudi Arabian complex is powered by 4,000 megawatts of solar, wind, and storage.[8]

The shocker is that would not even be close to the quantity of renewables needed just for that tiny amount of global air travel. The report's calculation covers only the renewables needed to produce green hydrogen. To power the direct air capture of CO_2 and the chemical synthesis of the e-fuels, you would need even more renewables—an equally staggering amount.

All proposed e-fuels—including ones for airplanes, ships, and cars—are prohibitively expensive, a major misuse of renewable energy, and some, like e-ammonia and e-methanol, raise huge environmental, safety, and health concerns. Indeed, a July 2024 Massachusetts Institute of Technology (MIT) study on maritime shipping found that "under current legislation, switching the global fleet to ammonia fuel could cause up to about 600,000 additional premature deaths each year."[9] Burning ammonia (NH_3) generates oxides of nitrogen, or NO_X (NO and NO_2), and "unburnt ammonia may slip out." NO_X and NH_3 can combine with other gases in the air to form fine particulates smaller than 2.5 microns (about 0.0001 inches) that have been found to cause heart attacks, strokes, and lung diseases such as asthma.

Ammonia combustion also generates nitrous oxide (N_2O), a potent GHG with 273 times the warming power of CO_2. A 2023 Princeton-led study looked at a scenario where e-ammonia fuel satisfies 5% of the world's energy demand and "1% of the nitrogen in that ammonia is lost as N_2O." Burning that ammonia as fuel could generate GHGs equivalent to 15% of today's fossil fuel emissions. And the GHG intensity of such combustion "would be more polluting than coal."[10]

We need a large amount of ammonia made from green hydrogen to directly replace the fossil fuel–based ammonia we use to make fertilizer. But like hydrogen, although ammonia is a useful chemical *feedstock*, it is a terrible *energy carrier*, as Paul Martin, a member of the Hydrogen Science Coalition, an expert group "seeking to bring an evidence-based view to hydrogen's role in the energy transition,"[11] spelled out in a detailed analysis. Martin, a chemical engineer who has worked with hydrogen and syngas for three decades, concluded that "anyone planning to either a) burn ammonia as a fuel or b) thermally crack ammonia back into hydrogen and nitrogen, *is an energetic and environmental vandal.*"[12]

In a May 2023 article on green shipping fuel, Reuters noted that "engines that can run on ammonia aren't yet commercially available."[13] The Boston Consulting Group explained in a 2023 report on hydrogen and e-fuels that electrolyzers used for such production "have a cost-overrun potential exceeding 500%."[14] We've already seen electrolyzer prices jump 40% to 50% between 2022 and 2023.

Another widely touted toxic e-fuel is e-methanol. Methanol (CH_3OH), also called wood alcohol, "is generally produced by steam-reforming natural gas to create a synthesis gas," which is in turn converted to methanol, the US Department of Energy explains.[15] Of the 110 million metric tons currently produced, less than 0.2 metric tons are from renewable sources.[16] In addition, methanol is uniquely toxic in very small doses; consuming as little as 4 milliliters can lead to vision loss. In an article on methanol toxicity, the online magazine *Autoevolution* concluded, "If we are to use renewable fuels, methanol should not be on the table."[17]

As Canary Media reported in April 2023, "Experts say that even if methanol producers can meet rising demand, the chemical is only expected to play a modest role in decarbonizing the shipping industry."[18] A September 2023 study in the *International Journal of Hydrogen Energy* put the price of e-methanol from green hydrogen and DAC at about ten times that of methanol from natural gas or coal.[19] Such a price differential suggests this pathway is unlikely to be a winning climate solution before 2050, particularly because no one has demonstrated that either hydrogen or DAC that is truly green can be commercial and scalable. As we'll see, trying to scale up e-methanol using CO_2 from bioenergy facilities is equally problematic.

Yet the hype is huge. A June 2023 article in *BBC Science Focus* magazine, "New E-Fuels Could Help Save the Planet (and Your Car) but at a Hidden Cost," managed to conclude, "E-fuels are a good idea."[20] Porsche has bet more than $150 million on e-fuels, and in January 2024, Dr. Karl Dums, the company's senior project lead for e-fuels, said, "I am absolutely confident we can save the world."[21] That same month, Canary Media wrote, "Ammonia, meanwhile, is predicted to become the leading fuel source for big ships in the long run."[22]

Wood Mackenzie, which calls itself "the leading global data and analytics solutions provider for renewables, energy and natural resources," published a June 2024 report asking, "Are synthetic fuels the key to unlocking growth in hydrogen?" The answer they provided: "The energy transition requires e-fuels."[23] In July 2024, Bloomberg reported, "European Commission President Ursula von der Leyen has said she will push for an exemption for cars running on so-called e-fuels under the bloc's plans to effectively end sales of combustion engine cars by 2035."[24] Such is the detrimental power of a bad idea.

Even if we had a simple solution for creating the green hydrogen needed for e-fuels, the technology would face huge barriers. The rest of this chapter will focus on why the other component of e-fuels such as e-kerosene and e-methanol—CO_2 pulled out of the air—is a counterproductive and unscalable climate strategy through 2050. The two most widely discussed ways of generating supposedly carbon-neutral CO_2 are DAC and bioenergy with carbon capture. Let's start by taking a closer look at DAC, which has been receiving far more attention than the bioenergy route.

DAC's Major Flaw Is the Same as Hydrogen's

Traditional carbon capture and storage systems recover and bury CO_2 from the exhaust gases of industrial facilities—particularly power plants—which have high concentrations of CO_2. In DAC, by contrast, enormous fans typically push large volumes of air over a liquid or solid sorbent that absorbs CO_2. Large volumes of air are needed because CO_2 in the air is so diluted. The Houston Astrodome, for example, contains only about 1 ton of CO_2, as CleanTechnica's Michael Barnard has explained.[25] Then, a great deal of energy is needed to release the CO_2 from the sorbents.

Like green hydrogen generation, the overall efficiency of DAC is quite low. In a 2019 report on negative emission technologies (NETs), a committee of the National Academy of Sciences calculated that the efficiency of a solid sorbent system is 7.6%–11.4% and of a liquid system is 4.1%–6.2%.[26] As with hydrogen generation, inefficiency is a key reason why the costs and energy consumption for DAC are so high. After a series of public workshops and meetings, the report explained, "The committee repeatedly encountered the view that NETs will primarily be deployed to reduce atmospheric CO_2 after fossil emissions are reduced to near zero."

DAC, like hydrogen, has a huge opportunity cost problem. A 2020 article that examined more than 200 studies, reports, and literature reviews on CO_2 removal and storage concluded that "the major, but generally ignored, policy issue about subsidizing renewables-powered DAC is whether renewable energy should be channeled for carbon removal rather than used directly to reduce carbon emissions by powering homes, industry, businesses, and transport."[27]

More recently, a number of studies have started to look at the opportunity cost issue and concluded that scaling up DAC won't make sense for a very long time. "Only when the region's electricity system is nearly completely decarbonized, do the opportunity costs of dedicating a low-carbon electricity source to DAC disappear," explained a 2021 analysis.[28] As one of the co-authors of that study, Princeton University professor Robert Socolow, told me in January 2024, the conclusion is broader. The opportunity costs for dedicating low-carbon electricity to DAC disappear only when the region's electricity system is nearly completely decarbonized *and* major sectors of the regional economy have been electrified and decarbonized.

The basic reason the opportunity cost is so high was explained in a 2022 review by the European Academies Science Advisory Council, the "collective voice of European science." They noted that "up to 20 times as much energy is required to remove a tonne of CO_2 from the atmosphere than to prevent that tonne entering in the first place."[29]

So we need to devote our time, energy, and resources to preventing CO_2 from being released into the air. For instance, using renewables to power electric vehicles (EVs) that replace ICE vehicles burning gasoline and diesel fuel is far more cost-effective at reducing CO_2 than using them to power DAC, as a 2021 analysis by the American Institute of Chemical Engineers concluded.[30] The analysis noted that EVs have "significant advantages over DAC"; for example, "The capital cost of DAC plants will be significant, while the cost of an EV is basically the difference in cost between an EV and an ICE" vehicle.

So EVs will have a far lower cost per ton of CO_2 reduced than DAC systems. Also, long before DAC starts significantly scaling up, EV costs will be equal to, if not lower than, those of ICEs, especially on a lifecycle basis. Indeed, studies suggest they are already close to parity.[31]

Until we have "a significant surplus of carbon-neutral/low-carbon energy," a 2020 article in *Nature Communications* warns, DAC is "unfortunately an energetically and financially costly distraction in effective mitigation of climate changes at a meaningful scale."[32] A 2020 review of carbon capture and storage CCS and direct air CCS reported that "according to one estimate, renewables-powered DAC would require all of the wind and solar energy generated in the U.S. in 2018 to capture just 1/10th of a Gt of CO_2."[33] That is 0.2% of global GHG emissions, or 1 part in 500.

Traditional CCS systems that recover and bury CO_2 from industrial facilities—particularly power plants—face significant difficulties in becoming profitable and efficient. The challenges DAC needs to overcome are even steeper. JP Morgan noted in its 2021 Annual Energy Paper that direct air carbon capture "energy needs appear to be 6x–10x higher than traditional CCS energy estimates, a process which itself is stuck in neutral."[34]

Just How Much Does DAC Cost?

"Capturing CO_2 from the air is the most expensive application of carbon capture," explained the International Energy Agency (IEA) in its 2022 DAC report.[35] The Congressional Research Service noted in 2021 that "generally, the more dilute the concentration of CO_2, the higher the cost to extract it, because much larger volumes are required to be processed."[36] Costs range from \$15–\$25 per ton of CO_2 for processes that produce "pure" or very concentrated CO_2 streams (such as fermenting corn ethanol) to \$40–\$120 per ton of CO_2 for processes with "dilute" gas streams, such as power generation, cement, and steel.[37]

DAC is the most expensive—currently several hundred dollars per ton of CO_2—because CO_2 is so diffuse in the atmosphere, with a concentration of about 0.04%. In comparison, flue gas of a coal-fired power plant is about 14% CO_2, over 300 times greater.[38] A 2018 analysis of DAC concluded, "CO_2 separation from air is unable to economically compete with CCS."[39] For the foreseeable future, it makes much more sense to put CCS on an existing coal plant than to build a DAC system with CCS.

A leading direct air CCS company, Climeworks, has said it sells CO_2 removal for \$600 to \$1,200 a ton. At a June 2023 DAC Summit hosted by Climeworks, the company's co-CEO Jan Wurzbacher told the audience his prices in 2050 could stay as high as \$300.[40]

In 2020, Microsoft requested proposals for carbon dioxide removal projects, and in 2021 the tech giant published a report on the results.[41] It received proposals for 189 projects and chose to purchase from fifteen suppliers. Only one was DAC, from a Climeworks facility in Iceland. Microsoft concluded that "at more than 50 times the cost per metric ton of most natural climate solutions, long-term solutions [such as DAC] today are both limited in availability and practically cost prohibitive."

Other experts agree that DAC will remain costly for some time to come. A 2021 article tried to gauge future prices by soliciting the judgment of eighteen experts in DAC technologies and negative emissions. The study found that experts expected the cost to decline "but to remain expensive (median by mid-century: around 200 USD/t CO_2)."[42] In a 2022 book chapter, MIT CCS expert Howard Herzog argued that the price of DAC in 2030 will probably not be below $600 per net ton of CO_2 removed.[43] Most studies report costs as dollars per gross ton of CO_2 removed, whereas in the real world, you must subtract out the CO_2 emissions created by the energy used to build and power the DAC system.

Most DAC systems will face enormous challenges running solely on zero-carbon power by 2030. New nuclear plants are far too expensive and likely to stay that way for a very long time (see Chapter 4). Truly green power must come from new renewables that are local and hourly matched, whereas a 2023 analysis found that "pairing [direct air capture and storage] to intermittent renewables is expensive."[44] But if the DAC system were powered by natural gas, the costs for the negative emissions could increase "by 250%, making it impractical." Herzog concludes that "the low range of cost estimates in the literature, $100–300/t$CO_2$, will not be reached anytime soon, if at all."

Although the cost for DAC systems might see steady and significant declines as the technology matures, this isn't a sure thing. A 2019 analysis of advanced fossil fuel plants pointed out that "cost estimates for early-stage technologies tend to be optimistic and poorly predict the actual cost of those technologies that reach commercialization."[45] Some expect DAC to decline in price rapidly as more plants are built, economies of scale are achieved, and people gain experience operating them. But as we've seen, the price of new nuclear power in the United States and Europe has risen for decades, and new small nuclear reactors are seeing an even higher price per megawatt.

Indeed, it's now clear that for energy systems, the more complex the design and the more they needed to be adapted and customized to their specific use environments, the slower the rate of cost decrease as sales volume increased.[46] Solar cells are both technologically simple and easy to standardize. That's a key reason prices have been dropping so sharply for so long.

But like nuclear power, a DAC system running on truly green power has both high system complexity and high customization for each deployment. So it is likely to see a slow rate of price decline. In the case of CCS, studies have shown that "there has not been a lot of interproject learning."[47]

Now imagine combining three completely different systems run on truly green power: electrolysis to make green hydrogen, DAC, and synthesis of hydrogen plus CO_2 into e-fuels. That is a high-complexity system requiring high customization. We've already seen that "electrolyzer projects tend to be highly complex, bespoke, and are proving far harder to construct than initially anticipated."[48] Systems to pull nitrogen out of the air to make e-ammonia are also unlikely to see rapid cost drops.[49]

So the three-part e-fuel system is unlikely to experience rapid and steady price declines. It is, at best, a post-2050 strategy if (1) every other sector of the economy has been decarbonized, (2) no other solution has been developed for this particular application of e-fuels, and (3) simply using direct air carbon capture and storage instead to achieve the same emission production is not a more viable option. Thus, although it's important to keep doing research and development on e-fuels, it is premature to consider scaling up any of them.

Bioenergy with Carbon Capture

Biomass can remove CO_2 from the air as it grows, and a carbon capture system on a bioenergy power plant can collect CO_2 that would—in theory—be carbon neutral. If it were carbon neutral, this CO_2 could be used with green hydrogen to make e-fuels.

But a growing body of research casts doubt on whether bioenergy and captured CO_2 from bioenergy are scalable climate solutions—or solutions at all. Those doubts are reinforced by findings from the first dynamic, integrated global modeling of bioenergy by the researchers of Climate Interactive in 2023:

- Policies to scale up bioenergy and bioenergy with carbon capture would *increase* global warming for several decades, with net cooling not occurring until 2100 or beyond.
- Scaling up bioenergy with carbon capture to 2–3 gigatons of CO_2 per year would require a land area the size of India.
- The best bioenergy strategy right now is to let bioenergy plants retire without replacement rather than install carbon capture systems on them.

So treating bioenergy as inherently zero-CO_2 or carbon neutral will invariably lead to policies harming the climate. This is already the case in Europe, where wood supplies most of the new biomass for bioenergy.

The world's top mitigation experts of the Intergovernmental Panel on Climate Change scaled back projections for bioenergy with carbon capture, utilization, and storage (BECCS) in their 2022 report, explaining that it "is not projected to be widely implemented for several decades."[50] The IEA has also steadily scaled back its use from "almost 5 Gt CO_2"[51]

removal by 2060 in 2017 to only 1 gigaton CO_2 removal a year by 2050 in its September 2023 1.5°C (2.7°F) scenario.[52]

Here's why BECCS is far less promising than originally thought.

We Are Increasingly Burning Trees for Biopower

Biomass is organic material, typically from trees, food crops (such as corn), dedicated energy crops (such as switchgrass), and waste from agriculture, forests, and mills. These other forms of biomass have similar emissions as wood, but wood, usually in pellet form, is the biomass of choice in places where biomass power is growing rapidly, such as the European Union.[53]

In its 2022 article "Europe Is Sacrificing Its Ancient Forests for Energy," the *New York Times* reported that "wood is now Europe's largest renewable energy source, far ahead of wind and solar."[54] The article noted that EU officials could not determine where 120 million metric tons of wood Europe used in 2021 came from, even though that's "a gap bigger than the size of Finland's entire timber industry." Forests in Finland and Estonia had been seen as sinks that pulled CO_2 from the air, but they "are now the source of so much logging that government scientists consider them carbon emitters."

Studies treating bioenergy as low carbon or carbon neutral tend to assume the "fuels are agricultural or forestry residues" but most of them aren't, explained one 2018 analysis.[55] Similarly, a significant number of business plans and policy analyses assume their company or country will just use residues. Yet as a 2019 review article noted, "The EU's own analyses[56] found that the amounts of residues available are insufficient (or already used in the forestry supply chain)" to satisfy the growing demand from large pellet producers—and that the primary source of biomass for US pellet producers was stemwood from trees.[57] Even if residues are used, burning wood residues is far from carbon neutral.[58]

The fact that we are increasingly burning trees is central to understanding why scaling up bioenergy would speed up global warming this century. That's because burning trees for BECCS results in a burst of "early-on positive CO_2 emissions" after the trees are cut down, as the Intergovernmental Panel on Climate Change wrote in 2022, and "net-negative effects are only achieved in time (carbon debt)" as the replanted seedlings grow and absorb CO_2 from the air.[59]

The carbon debt for biomass power is the total lifecycle emissions—from harvesting, drying, transporting, and burning the biomass—plus the forgone carbon removal, which is the loss of the future CO_2 that would have been captured if you hadn't cut the trees down. "If the forest had not been cut, it would have continued to grow, removing additional carbon from the atmosphere," explains a 2022 study.[60]

Although adding a carbon capture system to a biopower plant reduces the emissions from burning the biomass, it does not capture 100% of smokestack emissions. Carbon capture systems have been far less than 90% effective in the real world, as we saw in Chapter 5. Also, the carbon capture system consumes a great deal of energy whose emissions must also be accounted for in any lifecycle analysis.

It can take decades for the replanted seedlings to grow and absorb enough CO_2 to pay off this debt and make total lifecycle emissions for the system negative, compared with just leaving the original trees alone while deploying the best low-carbon alternatives to bioenergy at that time. Also, each year that more trees are harvested for bioenergy, the carbon debt just keeps rising. The debt doesn't start getting paid off until the scaling up stops.

Carbon Debt and Long Payback

How important is it to consider carbon debt when weighing the climate impacts of cutting down trees for biomass power? As a 2021 forest bioenergy letter by 500 scientists makes clear, the debt is significant: "When wood is harvested and burned, much—and often more than half—of the live wood in trees harvested is typically lost in harvesting and processing before it can supply energy, adding carbon to the atmosphere without replacing fossil fuels." Another reason the debt is big and the payback is long is that wood burning is inefficient compared with fossil fuels. "Overall, *for each kilowatt hour of heat or electricity produced, using wood initially is likely to add two to three times as much carbon to the air as using fossil fuels.*"[61]

The carbon debt is reduced or paid back over time if an equivalent number of new seedlings are replanted to replace the harvested trees and successfully grow to maturity. "But regrowth takes time the world does not have to solve climate change," the letter explains. "As numerous studies have shown, this burning of wood will increase warming for decades to centuries. That is true even when the wood replaces coal, oil or natural gas." In its 2019 report on negative emission technologies, the National Academy of Sciences similarly notes that harvesting live trees "may take decades or centuries to recover their original biomass and reach carbon sequestration parity."[62]

The payback time is fastest if the new biomass plant displaces a coal plant. But very few countries in the world besides China and India are building many new coal plants. In the United States and many other countries, the plant that would be displaced—the next power plant that would have been built instead—is a mixture of natural gas and renewables. As the electric grid increasingly decarbonizes over the next decade, biopower will increasingly be displacing zero- or very-low-carbon technologies, so this type of debt reduction will shrink toward zero, and the payback time will increase beyond a century.

Because of the carbon debt, a biomass power plant with carbon capture does not start out as a net remover of CO_2. In fact, it starts with a big surge of CO_2. How long it takes for the plant to shift from a net emitter to a net reducer of CO_2 can be determined only by a dynamic analysis of its lifecycle over decades. Because so much of the original biomass carbon is not captured, the net emissions removed from the atmosphere by the plant can be substantially less than the gross emissions captured at the smokestack. One 2013 analysis examined the lifecycle "carbon losses" from bioenergy with carbon capture using switchgrass, an energy crop.[63] It found that for every 1 ton of CO_2 sequestered, emission leakage in the entire supply chain is 1.11 tons.

Putting a CCS system on a bioenergy plant by itself will result in major losses, as noted earlier. Compared with power plants without CCS, the postcombustion carbon capture and compression system requires much more power to run, which is called the energy penalty or parasitic load. It is estimated to be 20%–30% of a power plant's capacity.[64]

A 2021 Chatham House report notes that R&D trials at a British plant imply that the overall system efficiency could drop from 36% to 21% relative to the same plant without a CCS system.[65] This suggests that models most often used to examine the strategies needed to limit global warming have greatly underestimated the amount of biomass—and land, water, energy, and fertilizer—needed for a given amount of emission reduction.

Also, the report notes that facility-level "capture rates are often cited as being 90% or more" by developers, modelers, academics, and policymakers. But according to a 2022 analysis of thirteen real-world flagship CCS projects—representing 55% of CO_2 capture capacity worldwide—actual capture rates are often below 70%, and "failed/underperforming projects considerably outnumbered successful experiences."[66] Significantly, there is usually a trade-off between the capture rate and the efficiency of a BECCS plant. As the capture rate rises, the BECCS system must draw more and more power from the plant. A higher capture rate means a greater energy penalty, which means lower overall efficiency for the system.

Scaling Up Bioenergy Worsens Warming Past 2100

Perhaps the most important finding in both the literature and Climate Interactive modeling results from analyzing an effort to scale up bioenergy over decades, which is the most likely scenario if it is to become a major climate solution. In that situation, as the scientists' letter notes, the carbon debt "increases over time as more trees are harvested for continuing bioenergy use."

Growth in wood consumption "causes steady growth in atmospheric CO_2 because more CO_2 is added to the atmosphere every year in initial carbon debt than is paid back by regrowth, worsening global warming and climate change," as a 2018 analysis explains.[67] The total carbon debt does not start getting repaid until biomass harvesting stops scaling up. The study was led by John Sterman, director of the System Dynamics Group at MIT's Sloan School who also helps lead the work at Climate Interactive. The study found that "growth in the wood pellet industry to displace coal aggravates global warming at least through the end of this century, even if the industry stops growing by 2050."

So scaling up wood-based bioenergy drives "an increase in CO_2, worsening global warming over the critical period through 2100 even if the wood offsets coal, the most carbon-intensive fossil fuel, the study concludes." But in reality, in most places, including the United States and European Union, the wood will be offsetting a far lower level of emissions. Indeed, by the time bioenergy with carbon capture could plausibly be ready to scale up rapidly, its primary competitors will be very-low-carbon technologies, such as renewables with longer-term storage. Ultimately, the carbon debt may never be repaid.

In 2021, one of the only dynamic forest-level analyses looked at a typical supply chain: pinewood from the US Southeast used to make pellets burned as fuel in the United Kingdom.[68] It found that after twenty years, "the uncaptured emissions from BECCS are equal to about 80% of what comes out of a coal plant's smokestack per megawatt-hour."[69] Bioenergy with carbon capture has high emissions, whether you are clear-cutting or just thinning a forest. The problem is that, twenty years from now, we can't be generating electricity that is only slightly cleaner than coal if we want to have any chance of meeting the Paris Agreement temperature targets and avoiding catastrophic climate change. We need to be generating electricity that is essentially carbon free.

The modelers at Climate Interactive took the dynamic analysis to a global level by using En-ROADS, which is "a global climate simulator that allows users to explore the impact that dozens of policies"—such as subsidizing CCS, pricing carbon, and improving agricultural practices—have on dozens and dozens of factors such as global temperature, carbon stored in forests, land usage, and CO_2 emissions.[70]

This modeling assumes that bioenergy with carbon capture proves to be both technically scalable, which has not been demonstrated yet, and commercially viable, for which the support of capital, citizens, and politicians remains untested. En-ROADS is a dynamic analysis (over decades) of the net system-wide CO_2 emissions removed from (or added to) the atmosphere, rather than a static snapshot of the gross CO_2 emissions captured from the smokestack. It is integrated and global, to examine key trade-offs such as using land for bioenergy versus land for food. Incorporating these trade-offs is essential because any model of bioenergy that effectively assumes we have a nearly limitless amount of land to devote to bioenergy is simply unrealistic.

Using En-ROADS, Climate Interactive found that policies to scale up biomass power with carbon capture would increase global temperatures for decades, with net cooling not occurring until the end of the century, even under optimistic assumptions.[71] As bioenergy (with or without carbon capture) is scaled higher, the temperature rise is also higher and lasts longer, well past 2100.

These findings are consistent with the recent scientific literature.

Consider a 2022 review by the European Academies' Science Advisory Council—the "collective voice of European science"—of the latest evidence on bioenergy. It "finds that there are substantial risks of it failing to achieve net removals at all, or that any removals are delayed beyond the critical period during which the world is seeking to meet Paris Agreement targets to limit warming to 1.5–2°C."[72] The review concludes that any bioenergy with carbon capture and storage "should be of limited scale, all feedstocks provided locally with very low supply chain emissions, and feedstock payback times should be very short." Such rules would eliminate virtually all current and planned bioenergy plants.

Bioenergy and Food Compete for Land

Even if bioenergy with carbon capture did generate significant net carbon neutral CO_2, it faces a tremendous obstacle to scalability because of the amount of land (and energy and water and fertilizer) required, even as the world faces "looming global land scarcity."[73] A study in *Energy Policy* notes that the land use changes needed for harvesting biomass "have contributed around one third of the world's cumulative CO_2 emissions since 1750."[74] Using satellite data, a landmark 2022 study in *Nature Food* found global cropland expansion has accelerated in this century, adding 250 million acres since 2000.[75] Because of the projected increase in food demand, "the vast majority of models estimate expansion of agricultural land by 2050, including several by more than half a billion hectares" (over a billion acres).

The 2022 *Land Gap Report* added up all the national climate pledges and found that more than 1.5 billion acres of reforestation is needed "to achieve the projected carbon removal, with the potential to displace food production including sustainable livelihoods for many smallholder farmers."[76] That means food production and reforestation could together require more land than the contiguous United States.

In 2011, I wrote a *Nature* article on "dust-bowlification and its potentially devastating impact on food security," explaining why feeding the world "by mid-century in the face of a rapidly worsening climate may be the greatest challenge the human race has ever faced."[77] Wildfires, sea level rise and saltwater intrusion, rising temperatures, extreme weather, and megadroughts threaten existing crops and farmland and are projected to generate tens of millions of climate refugees in the coming decades. They will all need new land to live on.

So how much new land will be left for new large-scale biomass production in the future? Not much. A 2021 synthesis report by the UN Food and Agriculture Organization concludes, "Land and water systems are at breaking point," and "agricultural systems [are] breaking down."[78] As a result, "there is little room for expanding the area of productive land, yet more than 95% of food is grown on land." The 2019 report by the World Resources Institute, the World Bank, and the UN agreed with an earlier study that found that "unless food demand patterns change significantly, there seems to be little spare land for bioenergy developments without a reduction of food availability."[79] The 2019 report's authors noted, "or, we add, without adverse effects on climate from losses of terrestrial carbon."

The Hype About Corn Ethanol and Hydrogen

The trade-off between food and fuel is important to understand because right now a remarkable two fifths of the US corn and soybean crops are burned in engines.[80] Fermenting the edible or starchy part of corn to make ethanol (C_2H_6O) results in a nearly pure stream of CO_2, which has led even serious organizations such as the IEA to suggest that putting carbon capture systems on corn ethanol biorefineries would be a good source of carbon neutral CO_2 for e-fuels. It wouldn't be.

In its December 2023 e-fuel report, the IEA notes that a major expansion of e-fuels such as e-kerosene and e-methanol would mean "a massive increase" in demand for CO_2 from low-carbon sources. Over

2 megatons of biogenic CO_2 captured each year globally comes from bioethanol plants, most of which are in the United States.[81] "There exists a significant potential synergy with biofuels production," the IEA explains, because byproduct CO_2 from bioethanol is "among the cheapest (USD 20–30/t CO_2) sources."

But the IEA went astray when it added, "Moreover, coming from sustainable biogenic sources, they enable the production of low life cycle GHG emission e-fuels." Unfortunately, as many studies have documented, bioethanol from corn does not deserve the label "sustainable." It certainly does not enable low-GHG e-fuels.

Consider the 2015 *Science* article "Do Biofuel Policies Seek to Cut Emissions by Cutting Food?"[82] Researchers looked at three of the main models used to set US and EU government policies and found that they "in effect are relying on decreases in food consumption to generate GHG savings." Between a quarter and a half of "the net calories in corn or wheat diverted to ethanol are not replaced but instead come out of food and feed consumption." Had the models included converting forests or grassland to grow more crops, they would have to add the CO_2 released from the land use change, "reducing or negating the net offset from producing more crops."

So although a number of analyses suggest that corn ethanol is about 46% less carbon intensive than gasoline,[83] many "modeling studies analyzing the GHG implications of using crops for biofuels find *little or no GHG savings if they take account of the conversion to agriculture of forests and grasslands necessary to replace the forgone food production*," as a 2015 World Resources Institute paper notes. That report adds, "yet present biofuel policies not only allow but even encourage biofuels to use crops from existing croplands."[84]

A 2022 study examined such land use conversions and concluded that they "caused enough domestic land-use change emissions such that the carbon intensity of corn ethanol produced" under US renewable fuel mandates "is no less than gasoline and likely at least 24% higher."[85] The study concluded, "Our findings confirm that contemporary corn ethanol production is unlikely to contribute to climate change mitigation."

Adding a carbon capture system to an ethanol plant doesn't improve the picture much. The 2019 World Resources Institute and World Bank report notes that such a system "only captures one-third of the carbon released by the whole process and therefore does not make the production of ethanol beneficial." Indeed, even if corn ethanol with carbon capture were a little better than gasoline, that isn't the appropriate comparison for investments we should make over the next decade. The world is transitioning rapidly toward EVs running on electric grids decarbonizing more every year.

An October 2023 analysis made clear that, compared with investments in ethanol with carbon capture, investments in vehicle electrification plus renewables reduce more CO_2 at a lower price by far.[86] And the electrification strategy does so with the least air pollution and land use. That is, shifting money and resources *from* carbon capture equipment and pipelines for ethanol biorefineries to "wind and solar farms for powering BEVs [battery electric vehicles] will benefit the climate, health, and land use tremendously while saving consumers enormous sums of money." Thus CCS systems for US ethanol production—plans for over forty such systems have been announced since 2018[87]—may not even deliver fuel with much lower CO_2 emissions than gasoline. But they will in any case deliver a fuel with far more carbon pollution and air pollution than EVs running on a cleaner and cleaner grid. Biodiesel fuel is equally problematic.[88]

Many studies have found that new cropland has long come at the expense of forests and grasslands, both of which store a great deal of carbon. A 2010 study concluded, "Across the tropics, we find that between 1980 and 2000 more than 55% of new agricultural land came at the expense of intact forests, and another 28% came from disturbed forests."[89]

Most of the remaining expansion came from shrubland conversion. Very little comes from previously cleared lands. The researchers concluded, "Expansion of the global agricultural land base inevitably means clearing tropical forest and shrubland ecosystems."

Crops such as corn, wheat, and rice have consumed 250 million acres since 2000. As *Science* reported in 2021, "Half of the new fields have replaced forests and other natural ecosystems that stored large amounts of carbon," including savannas, thereby "accelerating climate change."[90] A 2020 *Nature Communications* article noted that many studies and federal reports documented a renewed "conversion of grasslands and other natural and semi-natural areas to row-crop production."[91] This started at a time in the 2000s that "coincided with periods of high commodity prices, rapid build-out of the biofuels industry, and reductions to the extent of federal land conservation programs." Making ethanol from corn was rising sharply, and croplands expanded by more than 1 million acres per year. The predominant crop planted on newly cultivated land was corn, and "yields of new croplands were 10.9% lower" than the national average for corn. New soy cropland yields were 8.4% lower than average.

So, we should not count on steady yield productivity improvements to limit the land impact of large-scale biomass expansion. The 2020 *Nature Communications* study found that crops are increasingly grown on lower-quality land in less suitable regions, "a dual setback to production gains from cropland expansion." The best cropland was already being used. The study noted that field-scale studies of US cropland expansion "reveal globally significant carbon emissions."[92]

Indeed, a 2019 study found that significant CO_2 was released in this country "where new croplands primarily replace grasslands."[93] Between 2008 and 2012, cropland expansion accounted for nearly 3% of US CO_2 emissions. Grassland conversion played the biggest role, "with more than 90% of these emissions originating" from stocks of soil organic carbon. The authors found that land use change emissions are larger than previously thought and "may represent a climate forcing that is effectively irreversible over human relevant timescales."

At the same time, the UN Food and Agriculture Organization reported in 2022 that, from 2000 to 2018, "almost 90% of global deforestation worldwide is due to agricultural expansion."[94] The main driver is cropland expansion, which causes nearly 50%.

Soil Organic Carbon and Grasslands

Whether biomass comprises trees or crops, any planting displacing grasslands causes serious climate harm. A 2018 study found that California grasslands are a more resilient carbon sink than forests in responding to twenty-first-century climate changes.[95] During wildfires, the carbon in trees is released to the atmosphere, whereas the carbon fixed in grasslands typically stays in the roots and soil. The study concludes that grasslands, including tree-sparse rangelands, appear to be more capable than trees of maintaining carbon sinks this century.

A 2021 *Nature* study looked at data from 108 experiments involving elevated CO_2 to see the impact on carbon storage in plants and soils.[96] The key finding was that "plants will likely play a far less significant role in drawing down carbon than previously predicted."[97] The lead author, Dr. César Terrer, noted, "When a plant dies, some of the carbon that accumulated in its biomass may return to the atmosphere. In soils, carbon can be stored for centuries or millennia." Soils store more carbon globally than all plant biomass combined. Terrer concluded, "It would be a mistake to plant trees in natural grassland and savanna ecosystems. Our results suggest these grassy ecosystems with very few trees are also important for storing carbon in soil."

We also shouldn't plant trees in permanently snow-covered areas in Alaska, Canada, the Nordic countries, and Russia. The dark forests would absorb more heat than the white snow does and thus "have a warming effect that exceeds the cooling effect of reducing GHGs," as the National Academy of Sciences explained in 2019.[98] Wildfire-prone areas, which are expanding due to climate change, are also unsuitable for new tree planting.

Any real-world proposal for scaling up CO_2 capture from bioenergy to make e-fuels should explain exactly where the several hundreds of millions of acres of trees or energy crops would be planted. Indeed, many studies have noted that the world does not have much unoccupied and unused spare land, and so the pursuit of large-scale BECCS will have serious equity and climate justice implications. "The world's forests and savannahs are not 'empty lands' available for conversion to cultivation—a myth debunked long ago," argued a 2021 literature review.[99] Similarly, the 2022 *Land Gap Report* analyzed the impact of climate mitigation and land use changes on Indigenous peoples and local communities: "The vast majority of lands and forests targeted by national and international pledges on climate change mitigation and forest restoration are neither unclaimed nor unused. They constitute the customary lands and territories of indigenous peoples and local communities."[100]

The report explained that, generally, national climate pledges "have paid little attention to who, in practice, is living on, using and managing the lands involved, much less their existing land rights." It concluded, "Without a social justice lens, any attempt to fulfill the many land-based climate pledges is likely to perpetuate injustices."

Not the Best Use of Biomass

Bioenergy is not a scalable climate solution. So it's no surprise that a January 2020 study on "The Future of Bioenergy" had two key conclusions.[101] First, "the scale of bioenergy that both provides net climate benefits and can be sustainably produced is more limited than most models and scenarios predict." Second, it is time for policymakers to "limit near-term incentives for land-intensive bioenergy."

A 2012 study found that the climate benefits claimed for bioenergy are based mainly on "an accounting error that treats biomass as automatically 'carbon free,' meaning it counts the benefit of using land or biomass for energy without counting the cost of not using them for other purposes."[102]

Bioenergy with or without carbon capture has a major opportunity cost, since it's not the best use for either land or biomass if the goal is to cut CO_2. The 2020 study on "The Future of Bioenergy" noted that the electricity that can be generated from land use solar panels "is at least 50–100 times that from biomass."[103] Photosynthesis is an inefficient way to convert sunlight into power. The study concluded, "When biomass is available as a waste product or as a result of good stewardship practices, the best use of the material is for long-term storage as, for example, in the construction of buildings."

A 2021 report by Material Economics, co-funded by the European Union, concluded that biomass resources generally have "the highest value in a net-zero context" when they are used to produce bio-based materials, including "wood products, paper and board, textiles, and chemicals."[104] At the same time, demand for such applications "is expected to grow in the EU to replace more carbon-intensive materials, such as cement and steel in construction or plastics in packaging."

The bottom line is "an increase of demand for biomaterials on the order of 50% thus needs to be accounted for" by 2050. But since the priority use of biomass in midcentury will be feeding about 10 billion people in a climate-ravaged world, unlimited biomass won't be available for all purposes.

Biomaterials and reforestation are priority uses for biomass after food. The best evidence finds that increasing biomass for bioenergy is counterproductive from a climate perspective—and that using it to generate CO_2 for e-fuels is probably the most counterproductive idea of all. If we follow the research instead of the hype, hydrogen will not—and should not—create a significant new demand for biomass energy in the near future.

CHAPTER 7

Why Electric Vehicles Crushed Hydrogen Cars

"Despite the steady drumbeat of negative headlines about the electric vehicle market," Canary Media explained in May 2024, "EVs are breaking sales records across the globe."[1] In 2022, 10 million new plug-in hybrids and fully electric vehicles (EVs) were sold worldwide.

In 2023, 13.7 million were sold. In an April 2024 report, the International Energy Agency estimated that "electric car sales could reach around 17 million in 2024," which would be more than one fifth of total car sales.[2]

In contrast, global sales of hydrogen fuel cell cars fell by over 30% in 2023.[3] Global sales dropped from nearly 21,000 in 2022 to under 15,000 in 2023. A July 2024 story, "U.S. Hydrogen Car Sales Are Collapsing," noted that in the first half of the year, 322 hydrogen cars were sold, "82% lower than a year ago."[4]

Everyone who shares the goal of rapidly and cost-effectively removing CO_2 from our energy system—and who thinks hydrogen will play a major role in doing so—must face two fundamental questions.

First, if hydrogen is such a winning energy carrier, why doesn't any sector of the economy use it for that purpose after all these decades and billions of dollars in government subsidies and corporate investment? Again and again throughout this book, we've seen how high cost, inefficiency, safety concerns, the chicken-and-egg problem, and unexpectedly high greenhouse gas (GHG) emissions have limited hydrogen's growth.

Second, in the first head-to-head competition between energy carriers, why have electric cars crushed hydrogen cars by 1,000 to 1 in the marketplace?

This chapter will focus on the second question. It's worth understanding why hydrogen cars lost so badly to electric cars because the same essential factors that killed hydrogen cars also make it extremely unlikely hydrogen can compete in any sector where electricity can be used directly, including heavy trucking. They are why hydrogen is unlikely to make more than niche contributions to reducing CO_2 emissions as an energy carrier by 2050 and beyond.

Significantly, these factors were clear two decades ago.

The Seven Challenges Hydrogen (and All Alternative Fuels) Face

In the early 2000s, the threat of global warming was becoming clear, and the race was on to identify low-carbon alternatives to gasoline. Back then, the question was, What alternative fuel would be able to replace gasoline over the next two decades as we ramped up efforts to reverse the business-as-usual growth path in global warming pollution? We had to start making our vehicles far less polluting to avoid serious, if not catastrophic, climate change.

In 1992, Congress passed the Energy Policy Act, and president George H. W. Bush signed it into law. One of the goals was to reduce the amount of petroleum used for transportation by promoting the use of alternative fuels in cars and light trucks. These fuels included natural gas, methanol, ethanol, propane, electricity, and biodiesel. Alternative fuel vehicles

(AFVs) operate on these fuels, although many are dual-fueled. That is, they can also run on gasoline.

The Energy Policy Act established two goals in this area: having alternative fuels replace at least 10% of petroleum fuels in 2000 and at least 30% in 2010. Obviously, we did not come close to meeting those goals. Examining why will reveal all the flaws in hydrogen as an energy carrier because, as we'll see, trying to replace fossil fuels with green hydrogen in the economy is a very similar challenge to replacing gasoline with hydrogen (or any other fuel) in the transportation system.

The act mandated that some new vehicles purchased for state and federal government fleets must be AFVs. It also gave the Department of Energy (DOE) the authority to promote AFVs in various ways, such as by developing voluntary partnerships with city governments and others. The job of managing this work fell to the Office of Energy Efficiency and Renewable Energy, the office I helped oversee and ultimately run.

The initial results of this effort were documented in a 2000 report by the US Government Accountability Office (GAO). By 1999, some 1 million AFVs were on the road, only about 0.4% of all vehicles. In 1998, alternative fuels consumed in AFVs substituted for some 334 million gallons of gasoline, about 0.3% of that year's total consumption. An additional 3.9 billion gallons of corn ethanol and methanol replaced gasoline that year, but those fuels were blended with gasoline and consumed in conventional vehicles.

These results were far from a resounding success. My old office at the DOE is exceedingly good at helping to develop clean energy technologies such as solar and wind power—which today are the fastest-growing sources of carbon-free electricity in the world. We're also good at getting people to use more efficient versions of existing technology (more efficient lighting, motors, heating, and cooling). Our efforts were verified by the National Academy of Sciences as having saved businesses and consumers more than $30 billion in energy costs. But getting people to use AFVs proved to be exceedingly difficult. The GAO, which is often critical of the work of government agencies, did not criticize the DOE's actions. Instead, it concluded, "The goals in the act for fuel replacement are not being met principally because alternative fuel vehicles have significant economic disadvantages compared to conventional gasoline vehicles. Fundamental economic impediments—such as *the relatively low price of gasoline, the lack of refueling stations for alternative fuels, and the additional cost to purchase these vehicles*—explain much of why both mandated fleets and the general public are disinclined to acquire alternative fuel vehicles and use alternative fuels."

By 2003, a significant literature had emerged explaining this failure, like the GAO report. It identified seven major barriers to the success of alternative fuels such as hydrogen:

- High initial cost (of the vehicles)
- High fueling cost (compared with the alternative)
- Inconvenience of use (fueling time and range)
- Limited fueling infrastructure (chicken and egg problem)
- Safety and liability concerns
- Failure to provide cost-effective solutions to major energy and environmental problems
- Improvements in the competition

All AFVs that had been promoted with limited success by 2004—hydrogen vehicles, EVs, natural gas vehicles, methanol vehicles, and ethanol vehicles—suffered from several of these barriers. Any one of these barriers can be a showstopper for an AFV or an alternative fuel, even where other clear benefits are delivered. EVs deliver the clear benefit of zero tailpipe emissions and have lower per-mile costs than gasoline cars, but two decades ago issues such as range, refueling, and vehicle cost limited their success and caused most major auto companies to withdraw their EVs from the marketplace.

The limited fueling infrastructure has always been an intractable barrier, especially for hydrogen. In a 2008 article titled "The Car of the Perpetual Future," *The Economist* put it this way: "One thing holding back hydrogen vehicles is a chicken-and-egg problem: why build cars if there is nowhere to fill them up, or hydrogen filling-stations if there are no cars to use them? Just around the corner, honest."[5]

In the case of natural gas light-duty vehicles, the environmental benefits were oversold, as were the early cost estimates for both the vehicles and the refueling stations. "Early promoters often believe that 'prices just have to drop' and cited what turned out to be unachievable price levels," explained one 2002 study.[6] "*Those who relied on the early estimates had a sense of being abused.*" The study concluded, "Exaggerated claims have damaged the credibility of alternate transportation fuels, and have retarded acceptance, especially by large commercial purchasers."

Two decades ago, all AFVs suffered from a failure to provide cost-effective solutions to major energy and environmental problems. In 2003, the Department of Transportation Center for Climate Change and Environmental Forecasting released its analysis, *Fuel Options for Reducing Greenhouse Gas Emissions from Motor Vehicles.*[7] It assessed the potential for gasoline substitutes to reduce GHG emissions over the next twenty-five years. It concluded that "the reduction in GHG emissions from most gasoline substitutes would be modest" and that "promoting alternative fuels would be a costly strategy for reducing emissions."

One of the main reasons for this AFV failure is that the competition—gasoline-powered vehicles—had improved much more than many people had thought possible. Indeed, years earlier, when tailpipe emissions standards were being developed, few suspected that internal combustion engine vehicles running on reformulated gasoline could achieve the strict standard requiring no more than 0.02 grams per mile of NOx (nitrogen oxides), which help form tiny particulates that can penetrate into deep and sensitive parts of the lungs, leading to serious diseases.

But by 2003, a new generation of hybrid partial zero emission vehicles, such as the Toyota Prius and Ford Escape hybrid, substantially raised the bar for future AFVs. These vehicles have no chicken-and-egg problem (since they can be fueled everywhere), no different safety concerns than other gasoline cars, a substantially lower annual fuel bill, greater range, a 30%–50% reduction in GHG emissions, and a 90% reduction in tailpipe emissions. The vehicles cost a little more, but that was more than offset by the government incentive and the large reduction in gasoline costs, even ignoring the performance benefits. Compare that with many AFVs, whose environmental benefits, if any, typically come at the expense of a higher initial cost for the vehicle, a much higher annual fuel bill, a reduced range, and other undesirable attributes from the consumer's perspective.

Two decades ago, one decade ago, and today, hydrogen cars face enormous challenges in overcoming each of the seven major historical barriers to AFV success. For instance, South Korea saw a sharp decline in sales for hydrogen fuel cell vehicles (FCVs) since 2022—including a 67% drop from the first quarter of 2023 to the first quarter of 2024.[8] In May 2024, Korean analyst firm SNE Research explained the drop this way: "With the launch of next-generation [fuel cell EV] model delayed, what makes the situation worse are; the shortage of hydrogen charging infrastructure, defective hydrogen incidents, the increasing charging cost of hydrogen car, and unsolved issues with durability of hydrogen vehicle fuel cell. All of these negative factors have made customers turn away from hydrogen vehicles in the green car market."[9]

Let's look in more detail at how each of the seven barriers applies to hydrogen vehicles.

High Initial Cost for FCVs

The initial cost for hydrogen FCVs has been very high. Toyota's Mirai FCV started selling in Japan in December 2014—and in the United States in August 2015—for some $57,000 before incentives. But Pat Cox, former European Parliament president, said in a November 2014 talk on hydrogen cars that Toyota was probably losing between $66,000 and $133,000 on each car.

Ford Motor Company aggressively pursued FCVs in the first decade of this century. By 2009, their thirty Ford Focus FCVs had, combined, over a million miles on them. Yet after all this work, Ford's *Sustainability Report 2013/14* explained, "Even with the advances we have made in hydrogen technology over the past 10 years, we still have challenges to overcome before hydrogen FCVs can compete in the market with current vehicle technology. *The cost and durability of the fuel cell system are the most significant challenges. For example, extensive DOE analysis has not yet revealed an automotive fuel cell technology that meets the DOE's targets for real-world commercialization,* or that maintains proper performance throughout the targeted lifetime while staying within the targeted cost."[10]

In November 2023, Hydrogen Insight published a story headlined "Toyota's new hydrogen car to be significantly more expensive than its entry-level H2-powered Mirai."[11] With the kind of low global sales volumes FCVs are seeing, a manufacturer cannot achieve economies of scale and might not be able to even cover the cost of the auto manufacturing facility.

High Fueling Cost

The cost of hydrogen at the pump has always been challenging, especially since the goal is carbon-free hydrogen. Two decades ago, renewable energy was much more expensive than it is now, so using renewables to generate hydrogen by electrolyzing water (H_2O) was prohibitive. As discussed in Chapter 3, the only true very-low-CO_2 hydrogen is green hydrogen from renewable power sources that are new, local, and hourly matched. But this green hydrogen is still incredibly expensive, especially compared with using renewable power to directly power electric cars. A fundamental problem with delivering cheap, low-pollution hydrogen is the inability to solve the problem of how to build hydrogen fueling stations cheaply.

Between spring 2021 and summer 2023, True Zero—California's biggest hydrogen fuel retailer, which owns thirty-seven of the state's fifty-five hydrogen fueling stations—nearly tripled prices. *Hydrogen Insight* calculated in September 2023 that "it is now almost 14 times more expensive to drive a Toyota hydrogen car in California than a comparable Tesla EV."[12] A Toyota Mirai fueled with such hydrogen would cost about 50 cents a mile. That's over five times the cost of fueling a Prius. Bloomberg explained in April 2024 that filling up a hydrogen car costs California drivers "the equivalent of paying $14.60 for a gallon of gas."[13]

Significantly, in its December 2023 annual report on the state's hydrogen refueling system, the California Energy Commission and California Air Resources Board explained, "Most of the hydrogen dispensed in the California station network has been produced using fossil gas steam methane reformation."[14] So not only is the cost of hydrogen in the state incredibly high, but most of it is not even low carbon.

Two decades ago, John Heywood, former director of the Sloan Automotive Laboratory at the Massachusetts Institute of Technology, argued, "If the hydrogen does not come from renewable sources, then it is simply not worth doing, environmentally or economically."[15] True green hydrogen is expensive today because, as we've seen, the scientific literature makes clear that it must be made by electrolyzing water in new renewable power plants under strict conditions.

Today, producing and delivering affordable carbon-free hydrogen remains the sine qua non for it to play even a niche role in the fight to remove CO_2 from our energy system. But as previous chapters have made clear, we are not close to solving this problem.

Inconvenience of Use (Fueling Time and Range)

Over time, we have developed expectations about the ease and convenience of use of major consumer products. Vehicles with internal combustion engines running on liquid fossil fuels such as gasoline and diesel have dominated the marketplace for over a century. So we expect it to be easy and convenient to fuel our cars whenever we need to at one of the more than 145,000 fueling stations in the United States and perhaps a million or more total gasoline pumps.[16]

This convenience has three components. First, how long does it take to fuel the vehicle? Second, how far can the vehicle go on a fully fueled tank? Third, are fueling stations plentiful enough to eliminate the worry of being unable to find fuel when needed, especially on a trip? It has been argued that FCVs are better than EVs because they can be fueled more rapidly. But that comes with an additional barrier: fueling stations that are prohibitively expensive to build, especially because the fuel has to be very low carbon. And the larger tanks FCVs need come at the expense of passenger and storage space.

For a car to run on pure hydrogen, it must be able to store hydrogen safely, compactly, and cost-effectively on board, which is a major technical challenge. Hydrogen does have an exceptional energy content per unit mass, such as per pound or kilogram, nearly triple that of gasoline. However, the storage equipment on a car fitted for hydrogen use, such as pressurized tanks, adds significant weight to the system and negates this advantage.

Also, hydrogen has a far lower energy content per unit volume than gasoline. That is, hydrogen contains much less energy per gallon at the same pressure. At room temperature and pressure, hydrogen takes up about 3,000 times more space than gasoline containing an equivalent amount of energy.

To address this problem, compressed hydrogen has been used in demonstration vehicles for many years, and nearly all hydrogen cars use it today. Compressed hydrogen can be fueled quickly, and the tanks can be reused through many cycles without any deterioration of capacity or performance.

Hydrogen compression is a straightforward process using mature technology, and it is cheaper than liquefaction. In this process, hydrogen is compressed to 10,000 pounds per square inch (psi), compared with the atmospheric pressure of about 15 psi, for FCVs such as the Toyota Mirai. Increasing compression allows more fuel to be contained in a given volume and therefore increases the energy density. Even at these high pressures, however, hydrogen has far lower energy per unit volume than gasoline.

But compression to 10,000 psi is a multistage process that could require energy input equal to 15% of the hydrogen's usable energy content.[17] The main technical challenges associated with compressed hydrogen are the weight of the storage tank and the volume needed. Research focuses on developing systems that are stronger and lighter, able to contain hydrogen under extreme pressures, even in the event of a collision (safety issues are addressed later in this chapter). A 10,000-psi tank would take up seven to eight times the volume of an equivalent-energy gasoline tank.[18] Operation at such high pressures adds overall system complexity and requires more sophisticated materials and components, including seals and valves that are expensive.

Another drawback of compressed hydrogen tanks is that they are usually cylindrical to ensure integrity under pressure. This undermines flexibility in vehicle design, because when a large cylindrical tank is used, the vehicle must be designed around it. By contrast, tanks for liquid fossil fuels can be shaped according to the vehicle's needs. Also, using very high-pressure onboard storage would limit refueling options to places that invested in an expensive multistage compression and fueling system.

The cost of storage increases with the pressure attained. Some estimates have put the cost of a high-pressure storage vessel at more than 100 times the cost of a gasoline tank.[19] From a global warming perspective, compressing 1 kilogram of hydrogen into 10,000-psi onboard tanks could take 5 kilowatt-hours or more of energy. If that electricity does not come from new renewable power but instead comes from the overall electric grid—which still has many fossil fuel plants—then this hydrogen isn't green anymore.

Back in 2004, the National Academy of Sciences panel on hydrogen concluded, "The DOE should halt efforts on high-pressure tanks and cryogenic liquid storage. They have little promise of long-term practicality for light-duty vehicles."[20] Another 2004 study by the American Physical Society concluded that "a new material must be discovered" to solve the storage problem.[21] Ford explained in its 2013/2014 report, "Onboard hydrogen storage is another critical challenge to the commercial viability of hydrogen FCVs. Current demonstration vehicles use compressed gaseous hydrogen storage. However, the high-pressure tanks required for this storage use expensive reinforcement materials such as carbon fiber. In addition, the current tanks are large and difficult to package in a vehicle without unacceptable losses in passenger or cargo space."

Ford concluded that because of the "still significant challenges related to the cost and availability of hydrogen fuel and onboard hydrogen storage technology," we need "further scientific breakthroughs and continued engineering refinements" in order to "make fuel cell vehicle technology commercially viable."

A decade later, most fuel cell cars still use high-pressure tanks for onboard storage. As an alternative, many companies working on hydrogen-powered heavy-duty vehicles are considering liquid hydrogen storage.

Liquid hydrogen is widely used today for storing and transporting hydrogen (transporting it will be discussed in the next section). In general, liquids enjoy some advantages over gases from a storage and fueling perspective. Compared with gases, they have a high energy density and are easier to transport.

But hydrogen is anything but typical. At atmospheric pressure, hydrogen becomes a liquid only at the ultrafrigid temperature of –423°F (–253°C), just a few degrees above absolute zero, the coldest possible temperature. It can be stored only in a superinsulated tank, known as cryogenic storage. Refrigeration processes have inherent efficiency limitations, and hydrogen liquefaction generally requires multiple stages of compression and cooling.

So liquefying hydrogen requires expensive equipment and is very energy intensive, consuming as much as 40% of the energy in the hydrogen.[22] That's a key reason why vehicles using liquid hydrogen have such a low overall system efficiency and are unlikely to be a cost-effective climate solution. As the National Academy of Sciences said back in 2004, "The use of liquid hydrogen as fuel on board a light-duty passenger vehicle seems unlikely to meet the capacity and size requirements acceptable to the automotive industry. In addition, further obstacles to this approach include the high energy requirements for liquefying molecular hydrogen, safety concerns related to continuous hydrogen boil-off, and the escalating number of delivery trucks that would be on the road to meet demand in the middle years of scale-up."

Also, keeping the temperature that low in a tanker truck is challenging, especially because the liquid hydrogen sloshes back and forth, so it starts to boil off, dangerously increasing the pressure in the tank. This effect is especially problematic when a vehicle remains idle for a few days.

The excess hydrogen is typically vented into the air. But since hydrogen is such a potent heat-trapping gas, venting makes liquid hydrogen even less attractive in a world focused on reducing total warming.

Historically, tanks for FCVs have had a boil-off rate of up to 4% per day. Considerable effort has been going into reducing this rate, but even a daily boil-off rate of 0.4% would be problematic given just how potent hydrogen is at warming up the planet. Any analysis of the climate impacts of hydrogen solutions involving liquefaction will have to look at the full lifecycle leakage, including transporting the liquid hydrogen to fueling stations.

We have to transport the liquid hydrogen to the fueling station because trying to generate it onsite would be impractical and environmentally harmful. That's because the smaller the liquefaction plant, the more energy intensive the system is. Also, few fueling stations would have the space for such a plant, partly for safety reasons, as we'll see. In any case, you would need an impractically large amount of space for the renewable generation needed to keep this hydrogen green. Otherwise, we would be liquefying hydrogen by using power from the grid, which would make it quite dirty until the grid has eliminated virtually every fossil fuel plant.

The cost of cryogenic liquid storage—including the cost of the storage tank, the cost of the equipment to liquefy hydrogen, and the cost of electricity to run the equipment—remains high. Liquid hydrogen requires extreme precautions in handling because it is at such a low temperature. Fueling is typically done mechanically, such as with a robot arm—yet another added expense.

Whether hydrogen is stored as a gas or liquid, significant barriers keep it from becoming widely available and cost effective. So hydrogen fueling is not more convenient than simply charging up an EV, which can even be done at your home or, increasingly, everywhere from parking garages to shopping centers. This is especially true because of the great expense and logistical challenges of building out a large hydrogen fueling infrastructure.

Limited Fueling Infrastructure: Chicken-and-Egg Problem

Because hydrogen is such a deeply flawed energy carrier, there is no national or local hydrogen delivery infrastructure, like the one that exists for electricity down to the level of every building and home. For the same reason, there are no commercially successful technologies that run on hydrogen, the way there are thousands of incredibly successful products that run on electricity, such as cars, light bulbs, laptops, factories, and heat pumps.

This raises two fundamental questions. First, why would companies make massive investments in a national and local hydrogen delivery infrastructure until it's clear there is winning hydrogen-using technology to ensure they make their money back plus a profit? Remember, tens of billions of dollars of investment will be needed. A typical hydrogen fueling station costs about $2 million, so building just 25,000 would cost $50 billion.[23] Yet as of May 2024, the DOE reports there are 64,000 EV charging stations—with 175,000 total charging ports—and even that is widely seen as an insufficient number to sustain a growing EV market in this country. At the same time, there were fifty-six retail hydrogen fueling stations, of which fifty-five were in California and one was in Hawaii.[24]

Second, why would anyone invest in the vast manufacturing capacity needed to build enough affordable FCVs without a vast infrastructure to move hydrogen around cheaply and efficiently? Without volume sales of the vehicles to begin with, they will be prohibitively expensive, or the manufacturer will have to take large sustained losses—or both. But who will buy these vehicles if there isn't the fueling infrastructure available?

So who will go first and take the huge risk of stranded assets, making major investments that never pay for themselves, let alone make a profit? Will it be the fuel providers or the car companies?

This is the so-called chicken-and-egg problem. And the answer is that no one has gone first for two decades. Because the fueling stations are so expensive, federal and state governments can't afford to subsidize much buildout. Another key barrier was clear two decades ago: It's fossil fuel companies that have the most experience producing hydrogen and building the infrastructure for moving it around. But why would they build an expensive infrastructure for the purpose of competing with their existing products: gasoline and diesel fuel?

This intractable problem has never been solved for hydrogen, which is a major reason why there is neither a significant fueling infrastructure nor a major manufacturing capacity for vehicles.

Producing green hydrogen onsite at a fueling station will be impractical, at least until the grid itself is entirely zero carbon, as we've seen. There isn't space to have anywhere near enough renewables to make the necessary green hydrogen, and you'd probably need significant electricity storage capability onsite.

Since the hydrogen economy must be built around carbon-free hydrogen, the hydrogen must be transported to individual fueling stations from a central facility. We have three general methods of transporting hydrogen: specially insulated tanker trucks for liquid hydrogen, trucks carrying canisters of compressed hydrogen, or pipelines. Pipelines are certainly the best delivery means over longer distances, but as we'll see, it's implausible that they would be built any time soon, and in any case, they wouldn't be designed for local distribution.

Let's start with the first option: *tankers carrying liquid hydrogen.* As discussed in the previous section, they are far too impractical and unsustainable for large-scale distribution. Because liquefaction requires up to 40% of the energy content of the hydrogen, it takes an already expensive and inefficient process—renewable electricity powering an electrolyzer to make hydrogen, which in the car powers a fuel cell to make electricity again—and makes it even more expensive and inefficient.

There's been no breakthrough in hydrogen storage. Delivering large quantities of hydrogen by liquid hydrogen trailer makes less sense now that we know how much hydrogen can contribute to warming the planet. As the DOE explains, "Some amount of stored hydrogen will be lost through evaporation, or 'boil off' of liquefied hydrogen, especially when using small tanks with large surface-to-volume ratios."[25] Minimizing boil-off in tanker trucks is particularly challenging. The Hydrogen Fuel Cell Partnership, made up of "auto manufacturers, energy companies, fuel cell technology companies and government agencies," wrote in 2022, "Currently over 90% of small merchant hydrogen in North America is distributed via cryogenic liquid tanker truck. *Once stored, parasitic heat transfer results in boil-off losses typically between 7–40%.*"[26]

But to make hydrogen a plausible energy carrier today, we need virtually zero combined leakage and boil-off losses throughout the hydrogen production, delivery, and use infrastructure.[27] As noted earlier, hydrogen is an indirect GHG with thirty to forty times the heat-trapping ability of CO_2 over 20 years. In most potential applications, even small leaks and losses will negate whatever modest benefit hydrogen might have. Liquid tanker trucks just don't make sense as part of a climate solution.

Trailers carrying compressed hydrogen canisters are a flexible means of delivery best suited for early market introduction of hydrogen. But this is an expensive delivery method because hydrogen has such a low energy density. Even with high-pressure storage, very little hydrogen is actually being delivered.

Back in 2003, tube or canister trailers delivered less than 300 kilograms of hydrogen to a customer, perhaps enough to fully fill sixty fuel cell cars. An analysis by Ulf Bossel and Baldur Eliasson of hydrogen delivery in improved high-pressure canisters estimated that a 40-metric-ton truck would deliver only about 400 kilograms of hydrogen into onsite high-pressure storage.[28] From an energy perspective, this delivery approach is exceedingly wasteful. A 40,000-kilogram truck is being used to deliver a mere 400 kilograms of hydrogen. More than 39 unused metric tons, a tremendous amount of dead weight, would be hauled around our highways, burning up diesel fuel. Compare that with the fact that the same size truck carrying gasoline delivers some 26 metric tons of fuel (10,000 gallons), enough to fill perhaps 800 cars.

Although their analysis is more than twenty years old, Bossel and Eliasson draw two striking conclusions that are still relevant today. First, the energy consumed by tanker truck delivery itself can be a very high fraction of the total hydrogen energy delivered. For a delivery distance of 150 miles, the delivery energy equals nearly 20% of the usable energy in the hydrogen delivered. For 300 miles, the energy ratio approaches 40%.

Second, many trucks would be needed if they were the primary delivery means. They conclude, "It would take 15 tube-trailer hydrogen trucks to serve the same number of vehicles that are nowadays energized by a single 26-ton gasoline truck." They further note that "about one in 100 trucks is a gasoline or diesel tanker," or 1%, so replacing liquid fuels with hydrogen transported by tube truck could mean that more than 10% of the trucks on the road would be transporting hydrogen. This would raise serious logistical and safety concerns.

There has been no breakthrough in hydrogen storage in two decades. In 2022, hydrogen expert Michael Liebreich wrote, "Tube trailers carrying 750kg of compressed H_2 would require 16× more deliveries than the equivalent energy via diesel tankers."[29]

Finally, let's look at the final delivery option: *pipelines*. Today, several thousand miles of hydrogen pipelines are in use worldwide, of which several hundred miles are in the United States. They are short, operate at high pressure, and are located in industrial areas for large users.

Pipelines might be the least expensive option for delivering large quantities of hydrogen. They are the main option for transporting re-fined petroleum products across the country "because they are at least an order of magnitude cheaper than rail, barge, or road alternatives."[30] The United States has nearly 200,000 miles of interstate pipeline carry-ing petroleum in various forms.[31] We also have over 2,000,000 miles of interstate and local natural gas pipelines.

Hydrogen pipelines are very expensive in comparison to petroleum or natural gas pipelines, however, in part because they carry a fuel that is very diffuse and prone to leaks. Hydrogen is also highly reactive. It can cause many metals, including steel, to become brittle over time. A 2024 analysis found that large-scale interstate hydrogen pipelines (14 inches in diameter) had a capital cost of up to $7 million per mile, whereas smaller 8-inch pipelines for local distribution might cost half of that.[32]

Clearly, hydrogen infrastructure designers would need to be very stra-tegic about where pipelines are placed to minimize total construction costs. The problem is that it is far from clear where hydrogen will be produced and far from certain that hydrogen will succeed in the mar-ketplace. Siting major new oil and gas pipelines is politically and envi-ronmentally controversial, but in both cases, we know where the source of the oil and gas is, and typically we know with some confidence an endpoint with high demand for oil and gas, even if the path the pipeline might take between them is less certain.

We have very little idea today about which hydrogen generation pro-cesses will win in the marketplace over the next few decades. Nor can we know what political pressures will be applied to favor one generation of technology over another. In addition, as we've seen, pure hydrogen FCVs are uncompetitive with high-efficiency vehicles (hybrids and EVs).

All this uncertainty makes it unlikely that anyone would commit to spending billions of dollars on hydrogen pipelines before there are very high hydrogen flow rates with other means of transport and before the winners and losers have been determined at both the production and vehicle end of the marketplace. In short, for all their virtues, new pipelines will probably not be the main hydrogen transport means over at least the next two decades.

The Center for Strategic and International Studies noted in 2023, "Buildout of new pipelines is a capital- and time-intensive process that requires high and stable hydrogen demand, and the low molecular weight of hydrogen requires compressors to operate at three times the speed of that required for natural gas, further adding to operational costs."[33]

The challenge of building the pipelines is compounded by the fact that it is increasingly hard to site new pipelines in this country. Communities will not want a hydrogen pipeline running through them any more than they want a carbon dioxide pipeline.

Analyses of a future hydrogen economy tend to assume the end. And although pipelines are the desired end game and "the costs of a mature hydrogen pipeline system would be spread over many years," as the 2004 National Academy of Sciences panel noted, "the transition is difficult to imagine in detail."[34] The AFV problem is inherently a system problem where the transition issues are as much of the crux as the technological ones.

Finally, pipelines would not work for certain applications, such as delivering hydrogen to fueling stations for heavy-duty trucks running on liquid hydrogen. The stations would have to have expensive onsite electricity-intensive liquefaction capability—as well as contracts for a massive amount of new, local, and hourly matched renewables.

That means the station owners would have the inescapable chicken-and-egg problem: They would have to make this huge investment in a large number of stations before they knew that hydrogen-fueled heavy-duty trucks would be commercially viable *and* that heavy-duty electric trucks would not. Given the rapid and ongoing advances in EVs and batteries—and the many fatal limitations that hydrogen vehicles have relative to electric ones—that is almost certainly a losing bet. If so, each station would be stuck with a stranded asset in a marketplace awash with them. Indeed, the threat of that possibility alone will make it much harder to finance the infrastructure in the first place.

The conclusion I came to two decades ago was, "There do not seem to be any particularly attractive near-term options for delivering significant quantities of hydrogen to fueling stations."

Yet today, the prospects for hydrogen as an energy carrier that will significantly contribute to reducing GHGs by 2050 and beyond have decreased.

Hydrogen Safety

The challenge for any new fuel or technology is that people are used to the problems with the existing technology, but there's a spotlight on any problem with the new technology. For instance, studies by the Swedish Civil Contingencies Agency and EV FireSafe, a group funded by Australia's defense department, have found that the rate of fires per vehicle is over ten times higher for gasoline cars than for electric ones.[35] But gasoline car fires aren't news, whereas EV fires are because they are both new and different from gasoline fires.

Hydrogen will face this problem at an even greater level because it is such an unusual and hazardous element—everything about it is new and different. It doesn't help that people associate the infamous crash of the hydrogen-filled *Hindenburg* zeppelin with hydrogen. Exactly why the *Hindenburg* crashed remains a subject of debate, but the presence of hydrogen almost certainly played a big role.

Yet even with all the attention on hydrogen in recent years, a 2024 article in the *International Journal of Hydrogen Energy* on safety issues concluded, "The hidden hazards of hydrogen safety [are] not fully recognized."[36]

Hydrogen has major safety problems. They haven't changed in two decades, and they will be the same two decades from now because they are based on their unique physical properties. I discussed some of these briefly in Chapter 4, because the nuclear power industry has had to reckon with the risks posed by hydrogen.

But these dangers are far greater once hydrogen becomes widely introduced to the broader public outside a carefully controlled and regulated industrial setting—and into new settings where market and political dynamics may create pressure to weaken regulations and skimp on oversight and expensive safety practices that minimize dangerous and fatal accidents in industrial settings. That's especially true since, as the 2024 article on hydrogen safety concluded, "Human error is one of the most important causes of hydrogen accidents. Small human error could cause serious incidents."

Let's take a close look at the safety issue. Hydrogen is flammable over a far wider range of concentrations than natural gas and has a minimum ignition energy one twentieth that of gasoline. As explained in a 2002 report by Arthur D. Little Inc., "Operation of electronic devices (cell phones) can cause ignition" and "common static (sliding over a car seat) is about ten times what is needed to ignite hydrogen."[37] Even "electrical storms several miles away" can generate sufficient static electricity to ignite hydrogen, noted James Hansel, a senior safety engineer for Air Products and Chemicals Inc., a major global supplier of hydrogen, in a 1998 presentation to Ford Motor Company.[38]

At the same time, hydrogen is not just an invisible gas; it also burns invisibly. The flame speed is not only ten times faster than that of natural gas under normal circumstances, but as concentrations in the air rise to 10%, the speed can hit 1,000 feet per second—and at even higher concentrations, it can actually detonate and hit speeds faster than a mile per second.

Hence, leaks pose a significant fire hazard. At the same time, as we've seen, hydrogen, with its tiny molecule, is the most leak-prone gas, and in the case of liquid hydrogen storage, a fraction will boil off every day and may need to be vented. In industry, hydrogen is subject to strict and cumbersome codes and standards, especially when used in an enclosed space, where a leak might create a growing bubble of hydrogen gas. For example, fire codes and standards in the United States typically require large clearances between buildings using hydrogen and ones with combustible materials. Also, rooms with a significant risk of hydrogen leaks generally require massive ventilation to avoid the need for costly explosion-proof equipment.

"Fuel-cell vehicles will require modifications to garages, maintenance facilities, and on-road infrastructure [such as tunnels] that could be costly and difficult to implement," as the A.D. Little report noted. "Implementation of critical safety measures for closed public structures may pose a serious hurdle to widespread use of compressed hydrogen."

To allow us to easily detect leaks with our noses, odorants are added to flammable, odorless gases, as in the case of natural gas. But that is probably impractical with hydrogen. As Jim Campbell of Air Liquide explained, adding odorants to hydrogen "adds a contaminant that is poisonous to many fuel cell technologies," which "will add cost to the hydrogen energy picture, either in pretreatment to remove the contaminants, or in reduced service life of the affected systems."[39] Moreover, even if practical odorants could be found, they would be highly unlikely to leak and diffuse in tandem with hydrogen, the tiniest and lightest of all molecules.

The real danger exists that people might step into the nearly invisible hydrogen flame and burn themselves. According to James Hansel, "people have unknowingly walked into hydrogen flames."[40] How can invisible leaks and flames be detected? A variety of optical, ultraviolet, and infrared sensors are available, but Hansel notes that "fixed detectors indoors (e.g., inside some volume, compartment) or outdoors (e.g., no volume constraint), unless many are utilized, or the volume is small, *are not highly effective.*"

In fact, an Air Products safety document recommends, "If a leak is suspected in any part of a hydrogen system, it is good protection to hold a large square of paper before you and approach the suspected leak so that any invisible flame will strike the paper before striking you." NASA, which has decades of experience in dealing with hydrogen and access to the most sophisticated technology, also sees the limitation of sensors. In its safety guidelines for hydrogen systems, it lists four detection technologies. Three involve advanced sensors, but the fourth is decidedly low-tech: "A broom has been used for locating small hydrogen fires," since "a dry corn straw or sage grass broom easily ignites as it passes through a flame."[41]

Industrial codes and standards for hydrogen are thus strict for a good reason. Undetected hydrogen leaks cause some 22%–40% of hydrogen accidents.[42] And this is "despite the standard operating procedures, protective clothing, and electronic flame gas detectors provided to the limited number of hydrogen workers," as Russell Moy, former group leader for energy storage programs at Ford Motor Company, wrote in a 2003 article. Moy concludes, "With this track record, it is difficult to imagine how hydrogen risks can be managed acceptably by the general public when wide-scale deployment of the safety precautions would be costly and public compliance impossible to ensure."[43]

In addition, any high-pressure storage tank presents a risk of rupture, and, as we have seen, the current trend in hydrogen storage is to use very high pressures. Hydrogen pipelines are also likely to be under high pressure to keep the gas flowing. If a tank is punctured, it is critically important that the tank does not shatter and the fuel does not ignite. Storage tanks are routinely tested under a range of harsh conditions (including puncture by armor-piercing ammunition) to ensure safety. Still, gaseous hydrogen will leak through most materials over time. Therefore, storage tanks and pipelines for hydrogen must contain liners of a material (such as a polymer compound) that is impermeable to hydrogen.

Hydrogen is also highly reactive. It can cause many metals, including steel, to become brittle. This raises the risk of cracking or fissuring, resulting in "potentially catastrophic failure of pipelines." Even low-permeability liners will allow some hydrogen through to the tank material; therefore, storage tanks must be resistant to hydrogen embrittlement. Air Liquide's Jim Campbell noted that "higher strength materials are more susceptible to hydrogen embrittlement."[44]

Failure to Provide Cost-Effective Solutions to Major Energy and Environmental Problems

The central challenge for any AFV seeking government support beyond research and development is that deploying the AFVs and the infrastructure to support them must cost-effectively address some energy or environmental problem facing the nation. Yet as two hydrogen experts wrote two decades ago, "Hydrogen is neither the easiest nor the cheapest way to gain large near- and medium-term air pollution, greenhouse gas, or oil reduction benefits."[45] A 2004 analysis by the Pacific Northwest National Laboratory and the University of Maryland concluded that even "in the advanced technology case with a carbon constraint, hydrogen doesn't penetrate the transportation sector in a major way until after 2035."[46] A push to constrain carbon dioxide emissions actually delays the introduction of hydrogen cars because sources of zero-carbon hydrogen such as renewable power can achieve emission reductions far more cost-effectively simply by replacing planned or existing coal plants. As noted above, our efforts to reduce GHG emissions in the vehicle sector must not come at the expense of our efforts to reduce GHG emissions in the electric utility sector or other sectors such as heating for buildings and industry.

In fact, a January 2004 study by the European Commission Center for Joint Research concluded that using hydrogen as a transport fuel might well increase Europe's GHG emissions rather than reduce them.[47] That is because many pathways for making hydrogen, such as grid electrolysis, can be quite carbon intensive and because hydrogen fuel cells are so expensive that hydrogen internal combustion engine vehicles may be deployed instead.

When this study was released, using FCVs and hydrogen from zero-carbon sources such as renewable power or nuclear energy had a cost of avoided carbon dioxide of more than $600 per metric ton, which was more than a factor of ten higher than most other strategies being considered. Yet two decades later, a 2022 report by the Electric Power Research Institute found an almost identical result. That study found that if strict rules for green hydrogen were followed, the US Inflation Reduction Act tax credits for hydrogen would reduce CO_2 at a cost of $750 per ton—roughly ten times more than the other tax credits. If the strict rules weren't followed, emissions would actually increase, not decrease.

All analysis of AFVs should be conservative, stating clearly what is technologically and commercially possible today. When discussing the future, it should be equally clear that projections are speculative and will require both technology breakthroughs and major government intervention in the marketplace. Analysis should treat the likely competition fairly: If major advances in cost reduction and performance are projected for hydrogen technologies, similar advances should be projected for hybrids, batteries, and biofuels.

Hydrogen FCVs face major challenges to overcome all the barriers discussed here. In 2004, I noted, "It is possible we may never see a durable, affordable fuel cell vehicle with an efficiency, range, and annual fuel bill that matches even the best current hybrid vehicle." But the competition hydrogen faces is even tougher today, and it is likely we will never see a hydrogen-fueled vehicle that can come close to competing with a comparable EV.

Of all the AFVs and alternative fuels, FCVs running on hydrogen are probably the least likely to be a cost-effective solution to global warming.

Improvements in the Competition

Any effort to replace existing fuels or energy carriers with something completely new and still in development faces two key challenges. The first is to achieve all of its key price and performance targets—and to solve all of the practical real-world challenges—needed to become a commercial product. This can often take a decade or longer. In the case of nuclear power, for example, after several decades of trying, the problems have not been solved. So far, it hasn't happened for hydrogen energy, either.

The second challenge is that the competition is not sitting still during all this development time. As noted earlier, one of the reasons why alternative fuel cars failed to catch on in the 1980s, 1990s, and 2000s is that gasoline-powered cars showed that they could become ultra-low emission and much more fuel efficient, which raised the bar for all alternative fuels, including natural gas vehicles, hydrogen FCVs, and EVs.

What has proved fatal to hydrogen cars—the reason that they have failed in the marketplace and have no prospect of succeeding—is that their primary competition switched from gasoline cars to EVs.

As I argued two decades ago, if someone could build a viable electric car, no hydrogen car could compete with it. That was clear because in a head-to-head competition, electric cars would be superior in every respect to hydrogen cars: initial cost, fueling cost, ease and convenience of use, the extent of fueling infrastructure (chicken-and-egg problem), safety concerns, and the ability to cost-effectively address major energy and environmental problems.

The two inescapable advantages that cars running on electricity had over cars running on hydrogen—and still do today—are a much higher efficiency in using electricity (and hence renewable power) and an ability to solve the chicken-and-egg problem.

Using hydrogen made with renewable electricity (green hydrogen) as an energy carrier for a device that ultimately runs on electricity is inherently much more inefficient than simply using the electricity directly. That is because of the tremendous overall system inefficiency of generating hydrogen from electricity, transporting hydrogen, storing it onboard the vehicle, and then running it through the fuel cell, as I (and others) explained two decades ago. "The total well-to-wheels efficiency with which a hydrogen fuel cell vehicle might utilize renewable electricity is roughly 20%. . . . The well-to-wheels efficiency of charging an onboard battery and then discharging it to run an electric motor in a PHEV [plug-in hybrid EV], however, is 80% . . . four times more efficient than current hydrogen fuel cell vehicle pathways."

Ulf Bossel, founder of the European Fuel Cell Forum, wrote a 2004 article, "The Hydrogen 'Illusion': Why Electrons Are a Better Energy Carrier," that concluded, "The daily drive to work in a hydrogen fuel cell car will cost four times more than in an electric or hybrid vehicle."[48] Alec Brooks, who led the development of the Impact EV, also explained in 2004, "Fuel cell vehicles that operate on hydrogen made with electrolysis consume four times as much electricity per mile as similarly sized battery electric vehicles."[49]

From a policy perspective, replacing most US ground transport fuels (gasoline and diesel) by 2050 with green hydrogen would require over 1,000 gigawatts of additional renewable power plants built for the US grid compared with replacing those fuels with carbon-free power. That extra capacity is comparable to the entire US electric grid from all power sources today. It is one of the many fatal limitations of hydrogen as a major GHG mitigation strategy.

From a consumer perspective, this fourfold inefficiency—and four-fold higher fuel price—is just another one of the many fatal limitations of hydrogen. In the real world today, driving to work in a hydrogen fuel cell car in California costs fourteen times as much as driving an EV.

As discussed above, the chicken-and-egg problem for hydrogen is an equally fatal limitation. It has never been solved because hydrogen has never had any success as an energy carrier, so it's a big gamble to build a manufacturing capacity for a hydrogen-using product such as a car when there's no infrastructure to make the fuel and deliver it to those cars. But at the same time, there's no incentive for someone to build out that expensive infrastructure until it's clear that some hydrogen-using product is going to be successful in the marketplace. And after decades of failure to build such a product, that incentive is unlikely to change any time soon.

Electricity doesn't have any of these problems. It has had tremendous success as an energy carrier. Also, electricity generation is widespread, and electricity infrastructure is already ubiquitous: It is delivered everywhere in the country, all the way to the residential level. Electric charging stations are cheaper than hydrogen fueling stations by more than a factor of ten. While the fossil fuel industry has little incentive to build out an enormous and expensive hydrogen infrastructure that competes with its own liquid fuels, the electric utility industry has a big incentive to make charging ports ubiquitous. That's because EVs are an entirely new market and source of profits for utilities.

In 2004, it wasn't clear that a pure EV would succeed in the market-place any time soon. Tesla wasn't incorporated until July 2003, and it didn't produce its first car until 2008—the expensive high-end Roadster sports car. But when I updated the book for the 2005 paperback, it was clear that you could build a plug-in hybrid EV. I called these e-hybrids, but they quickly became plug-in hybrid EVs (PHEVs).

Because they have a gasoline engine and are thus a dual-fuel vehicle,

PHEVs avoid two of the biggest problems of pure EVs. First, they are not limited in range by the total amount of battery charge. If the initial battery charge runs low, the car can run purely on gasoline and on whatever charging is possible from the regenerative braking. Second, EVs take many hours to charge, so if, for some reason, owners were unable to allow the car to charge—either because they lacked the time between trips to charge or there was no local charging capability—then the pure electric car could not be driven. Thus, PHEVs combine the best of both hybrids and pure EVs.

PHEVs also do not have a high fueling cost compared with gasoline. In fact, the per-mile fueling cost of running on electricity is about one third of the per-mile cost of running on gasoline. That means the higher first cost would be partially and eventually fully compensated for by the economic benefit of a lower fuel bill.

Indeed, a number of companies did launch PHEVs around 2009, including General Motors with the Chevy Volt, Toyota with a Prius PHEV, the Chinese company BYD with its F3DM PHEV, and Ford with the Escape PHEV.[50] Today, a number of PHEVs are still popular, but the price and performance of electric batteries improved at such a surprising rate in the past two decades that it's clear all-electric vehicles are the inevitable winner.

Ultimately, there is one central lesson from EVs' complete domination of the AFV market and the failure to achieve even a moderate, thriving niche for hydrogen fuel cell cars. The electric system is all but certain to be the dominant winner in any sector with both a hydrogen solution and an electrification solution. This means that just as hydrogen FCVs will always be the car of the perpetual future, more and more of the "hydrogen economy" will be in the perpetual future.

False Promises and Real Solutions in the Race to Save the Climate

THE RACE TO SAVE THE CLIMATE motivated the first edition of this book in 2004, and it is the motivation for this major new update. Two decades ago, it was clear that climate change was an existential threat to human civilization and that hydrogen was unlikely to be a major contributor to addressing it by 2050.

Since then, global greenhouse gas (GHG) emissions have soared more than 40% to the equivalent of 50 billion tons of CO_2 a year, GHG emissions from hydrogen production have doubled, and scientists' warnings of what would happen if we didn't take strong action have come true. "The past 10 consecutive years have been the warmest 10 on record," as NASA reported in 2024.[1] The melting of Arctic sea ice and the great ice sheets of Greenland and Antarctica has actually been faster than expected, and some of the most dangerous forms of extreme weather—heat waves, hurricanes, deluges, droughts, and wildfires—have hit record proportions. Leading experts are "astonished."[2]

Scientists now warn us we're rapidly approaching temperatures that risk ever-worsening and ultimately catastrophic climate impacts that will be irreversible for generations. Based on policies currently in place globally, we are headed to as much as 6°F (3.3°C) warming by 2100. So we must rapidly implement much more powerful policies.

The science has kept growing stronger that the world needs to hold total warming to "well below 2°C" (3.6°F) above preindustrial levels while "pursuing efforts to limit the temperature increase to 1.5°C" (2.7°F), as the nations of the world agreed unanimously in the 2015 Paris Accord and have repeatedly reaffirmed in subsequent meetings. The world's top climate experts of the Intergovernmental Panel on Climate Change have made clear that keeping warming to 2.7°F requires "net zero" global emissions by 2050, and keeping emissions below 3.6°F requires very deep reductions by 2050 and net zero global emissions just a decade or two later.

"Net zero" means that whatever emissions the world can't mitigate would be offset by carbon dioxide removal (CDR) technologies that currently aren't commercial. But the two most studied and modeled CDR technologies are direct air capture and bioenergy with carbon capture and storage (CCS)—and we saw in Chapter 6 that these are not promising strategies to scale. That's why both the Intergovernmental Panel on Climate Change and the International Energy Agency (IEA) have sharply scaled back their projected use by midcentury.

If we don't "drastically reduce emissions first," then CDR "will be next to useless," argued CDR expert David T. Ho, an oceanography professor, in a 2023 *Nature* article.[3] Ho, who was a reviewer for the $100-million XPRIZE Carbon Removal competition, concluded, "We must be prepared for CDR to be a failure."

So prudent planning means we must strive to reduce global GHGs as close to actual zero by midcentury. That means our top focus must be deploying existing commercial zero-carbon technology far more rapidly than we are now. Technologies that need at least one breakthrough to become commercial and scalable, such as fusion or hydrogen as an energy carrier, are unlikely to significantly contribute to meeting global temperature targets.

Equally important, we must stop building infrastructure that is not zero carbon. Based on IEA analysis, their executive director Fatih Birol told the *Guardian* back in 2018, "We have no room to build anything that emits CO_2 emissions."[4]

So today, we know hydrogen is poised to be worse than a boondoggle for two reasons. First, scaling up a hydrogen economy would have a huge opportunity cost: diverting vast amounts of renewables and money away from the real solution, which is scaling up the electrification economy. Until the economy has reduced GHG emissions by more than 90%, the best use of both renewables and money is to directly replace the use of fossil fuels in power plants with renewable energy and advanced storage in order to power key commercial electric technologies that are rapidly capturing market share. These include electric vehicles and heat pumps—as well as emerging electrification technologies for building and industrial heating, heavy industry, and the like.

The Inflation Reduction Act by itself could potentially misallocate a half trillion dollars or more toward hydrogen and hydrogen-related technologies that—if green hydrogen production follows strict rules and drops in price—would achieve emission reductions at a cost of $750 per ton of CO_2, about ten times more than the other tax credits. But in the real world, few, if any, CO_2 reductions are likely to occur because those rules may be weakened or simply unenforced, and even tiny hydrogen leaks would wipe out the modest benefit from hydrogen.

Indeed, the second reason scaling up hydrogen is a looming climate disaster is that we now know it is an indirect GHG with a very high global warming potential. The latest studies find that hydrogen has thirty to forty times the warming power of CO_2 over a twenty-year period, the most critical period for reaching the goals of the Paris Agreement and avoiding catastrophic and irreversible climate impacts.

A 2023 *Nature* article explained, "By reducing short-lived climate forcers, such as hydrogen, there is potential to slow down global warming in the next 20–25 years."[5] But the nation and the world are planning to massively scale up hydrogen, which multiple studies conclude will inevitably lead to much higher emissions of this leakiest of gases.

Hydrogen is not only notoriously leaky but also odorless, invisible, and hard to detect. Undetected leaks cause up to 40% of all hydrogen accidents, partly because hydrogen burns invisibly, and fires are so easy to start they can even be ignited by static electricity. Worse, some of the most popular supposedly low-carbon ways of making and using hydrogen are especially leaky. A 2022 journal article on the subject concludes, "The supply chain route has a significant impact on emissions" of hydrogen (H_2) from leaks. "The green H_2 routes have higher emissions" (2.6% on average) compared with the blue and biomass routes (average 1.4%).[6]

A 2022 study from Columbia University's Center on Global Energy Policy notes that in 2050, "certain leaky processes will be more widely used (if they were used at all) than they were in 2020."[7] Therefore, they project a 2050 economy-wide leakage rate "between 2.9 percent (low-risk case) and 5.6 percent (high-risk case)."

These are huge leakage rates for gas with such high near-term warming impacts—especially since we are pursuing a rapid scale-up of hydrogen use, which means a rapid scale-up of hydrogen leaked into the atmosphere.

Liquefied hydrogen (LH_2), which is the cheapest and easiest way to truck the gas around, is doubly problematic from a climate perspective. Liquefaction to –423°F consumes up to 40% of the energy in the hydrogen, and it inherently leads to emissions from leaking and venting. As the 2022 journal article notes, "The LH_2 supply chains, particularly the ones which transport green H_2, were found to be the highest emitting."

How polluting is liquid hydrogen? In 2022, the European

Commission's Joint Research Centre released a summary report from an expert workshop on the potential global warming impact of hydrogen emissions. It cites Air Liquide's stunning estimate of 10%–20% leakage from the LH_2 supply chain.[8]

Remember, hydrogen doesn't have significant near-term or medium-term benefits as an energy carrier when compared with other options, especially electricity, but it does have a huge opportunity cost. Also, it isn't the inevitable winner in the race to save the climate but rather the likely loser in almost every sector. So investments in anything less than near-zero-GHG lifecycle emissions cannot be justified.

Hydrogen leakage is one of the many reasons heavy-duty trucks running on liquid hydrogen will have much higher GHG emissions than trucks running on electricity and could well have higher emissions than trucks running on diesel fuel. Leaks are also one of the many reasons why e-fuels are not a practical and scalable climate solution for ships or airplanes.

Some argue that e-ammonia (NH_3) is the inevitable winner for ships because it does not have any carbon to oxidize into CO_2 when burned. But it releases something much more potent: N_2O, which has 273 times the warming power of CO_2. A 2022 article in *Nature Energy* looks at the impact "if 0.4% of the nitrogen in ammonia fuel were to become N_2O, whether directly or indirectly"—an emission rate consistent with research on shipping emissions, the study notes.[9] Researchers found "these emissions would completely offset the GHG emissions benefits of switching fuels in the first place . . . even if the ammonia production and distribution produced zero emissions." But, of course, e-ammonia production doesn't produce zero emissions, as we've seen. Green hydrogen has significant emissions, including its own leakage problem. It is hard to imagine a plausible real-world scenario where e-ammonia has a significant net climate benefit for shipping by 2050.

Hydrogen will be a climate problem, not a climate solution, for the next two decades at least. Currently, over 98% of hydrogen is still made from fossil fuels for use as a chemical feedstock, primarily for petrochemicals and ammonia fertilizers. GHG emissions from that hydrogen production are comparable to those of aviation or shipping—and that's without factoring in hydrogen leakage and venting from the hundred million tons of hydrogen produced each year.

So our top hydrogen priority is replacing dirty hydrogen with green hydrogen from renewables that are new, local, and hourly matched. Just doing that will be much harder and more expensive than most people realize for several reasons. First, as noted earlier, such replacement would require generating as much renewable electricity as the United States produces each year from all sources including coal and gas plants, which in turn will require a staggering number of electrolyzers.

Unfortunately, electrolyzers "have a cost-overrun potential exceeding 500%,"[10] and prices jumped upward of 50% in 2023 because "electrolyzer projects tend to be highly complex, bespoke and are proving far harder to construct than initially anticipated."[11] Second, it is hard to find an optimistic study about hydrogen that does not simply assume rapid price drops for green hydrogen. Yet studies have found that major energy construction projects such as green hydrogen that are both highly complex and bespoke often see very slow price declines or, as in the case of nuclear power, sometimes even see a steady rise in prices. Indeed, a September 2024 BloombergNEF study predicted that electrolysis system capital costs will fall by only "about one-half from today to 2050 in China, Europe, and the US."[12]

Third, if we truly want green hydrogen, we must design electrolyzer systems, all interconnecting elements, and infrastructure with near-zero leakage and include monitoring and verification systems. Fourth, liquid hydrogen cannot plausibly be part of the storage or delivery system for the foreseeable future until we've designed virtually leakproof systems with no substantial boil-off and venting. It isn't easy being green hydrogen.

The head of EU lending at the European Investment Bank told Bloomberg in May 2024, "At the moment, there is no case that we can see for a model based on green hydrogen independently produced and distributed as an energy commodity."[13] Yet, as we've seen, all of the other types of hydrogen are much worse than green hydrogen. Gray hydrogen from natural gas and blue hydrogen from gas with CCS are much dirtier. Geologic hydrogen will also be dirtier and unscalable. Hydrogen from new nuclear plants will be even more expensive and raise significant safety problems.

Nonetheless, we need to address this large and growing climate problem from current uses of hydrogen, which is likely to take decades. Until then, it makes little sense to spend vast sums building entirely new manufacturing systems for ships, planes, and heavy-duty trucks based around hydrogen or e-fuels—and a comparable amount of money on building a new fuel delivery infrastructure.

It is entirely possible, if not probable, that this infrastructure, along with associated manufacturing systems, will become stranded investments, much like the billions of dollars that have been spent trying to build a manufacturing capability and fueling infrastructure for hydrogen cars. Two decades ago, we knew that was a lost cause, that hydrogen cars could never compete with electric vehicles for several reasons. And so it should surprise no one that electric cars are now crushing hydrogen cars by 1,000 to 1 in total sales.

After decades of trying, hydrogen is still used almost exclusively as a chemical feedstock. "The uptake of hydrogen in new applications in heavy industry, transport, the production of hydrogen-based fuels or electricity generation and storage . . . remains minimal, accounting for less than 0.1% of global demand," as the IEA reported in September 2023.[14]

The fact that every alternative fuel has several hurdles to jump is why they have such a high failure rate in the real world—and why the dream of a hydrogen economy has gone nowhere. Fundamentally, the seven major barriers to the success of alternative fuels for vehicles can be applied to other applications of hydrogen more broadly:

- High initial cost (of the hydrogen-using technology)
- High fueling cost (compared with the alternative)
- Inconvenience of use
- Limited fueling infrastructure (chicken-and-egg problem)
- Safety and liability concerns
- Failure to provide cost-effective solutions to climate change
- Improvements in the competition (electrification technologies)

Hydrogen is much more implausible as a climate solution today than two decades ago because few applications can overcome the crucial sixth barrier and because its primary competition—electrification—has improved more than almost anyone imagined back then, especially compared with hydrogen. Along with the harsh reality of hydrogen's failure, we should recognize the growth in electrification technologies as one reason for hope in the fight against climate change.

Electrification Is the Climate Solution

As noted, the core enabling technologies for electrification are advanced batteries, cheap solar and wind power, electric vehicles, and electric heat pumps. These have all made astonishing progress in the past two decades to become tremendous commercial successes. Batteries, wind, and solar, particularly, continue to see stunning price drops. Meanwhile, the core enabling technologies for hydrogen—especially fuel cells and hydrogen storage—have stagnated.

Two decades ago, we knew that hydrogen almost certainly could not compete head-to-head with a direct electrification solution because of the tremendous cost and inefficiency of turning renewable power into hydrogen, storing and transporting it, and then turning it back into electricity. What we didn't know then was how much of the fossil fuel economy beyond passenger vehicles could be electrified.

But now it seems that almost every month we hear of a company commercializing an electrification solution that competes with hydrogen. As IEA energy analyst Dr. Simon Bennett, who coauthored their 200-page *Future of Hydrogen* report in 2019, told me in August 2024, "Many of the technologies that have generated the most excitement in the past decade were ones that had the effect of outcompeting hydrogen or CCS in an application where we used to think hydrogen was the best approach."[15] Dr. Bennett, who previously worked as a programme manager at the European Commission with responsibility for CCS projects and innovation, said this pattern started with EVs and high-performance batteries. But now it has extended to electrification of steel manufacturing, nonhydrogen long-duration energy storage, heat pumps, limestone-free cement, and enhanced geothermal systems that may be able to provide flexible 24/7 power that is dispatchable to the grid when needed.

"The path of innovation is pointing in the direction of bypassing H_2 and H_2-related CCS applications over the longer term," explained Dr. Bennett. "With a slighter longer time horizon, we might expect innovation to deliver elegant non-hydrogen solutions. That could reduce some of the expectations for hydrogen. We should all be alive to how the risks of different strategies evolve and ensure we can adjust course as necessary."

A key point of this book is that we do not have to have zero GHG emissions in 2050. We need to get as close as we can as fast as possible without getting distracted by false promises. It would be a mistake to scale up strategies now that are unlikely to be the long-term winners but would misallocate vast amounts of money and renewables that could cut far more emissions at far lower costs over the next two decades. We do not want to end up with tens if not hundreds of billions of dollars in stranded assets in sectors of the economy with only a few percent of global emissions when we aren't even close to having done the much more straightforward job of reducing emissions 90% by deploying the technologies we know today are scalable marketplace winners.

Are there any sectors for which hydrogen is the truly best and only climate-friendly option? Heavy-duty trucks running on liquid hydrogen from green renewables do not appear to be able to beat even current fossil fuel–powered trucks on any of the seven crucial barriers. If the hydrogen isn't new, local, and hourly matched or if it is dirty blue hydrogen with CCS, the entire effort will only speed up warming. Given the rapid improvement in battery and electric vehicle technology, any hydrogen truck is a bad bet, but the worst is the liquid hydrogen version.

The natural gas industry has been pushing to use hydrogen to heat homes and other buildings, as *IEEE Spectrum* explained in 2022.[16] But as one European expert who reviewed thirty-two independent studies on that subject told them, "You need five to six times more electricity to generate the same amount of heat with hydrogen than with an efficient heat pump. That's the Achilles' heel for hydrogen. It's basic thermodynamics."

Hydrogen is often touted for industrial heat, which requires much higher temperatures than buildings. But as the nonprofit American Council for an Energy-Efficient Economy, which ran an industrial heat pump (IHP) workshop in October 2023, explains, "Commercially available electrically powered IHPs can provide process heat up to 160°C (320°F)."[17] That covers about a third of industrial heat. Heat pumps "currently in development" can supply heat up to 280°C (536°F), which would cover more than two fifths (40%) of industrial heat.

But beyond heat pumps, there are many other industrial electrification technologies. In fact, a November 2020 *Environmental Research Letters* article on industrial decarbonization in Europe details that "seventy-eight percent of the energy demand is electrifiable with technologies that are already established," and 99% is achievable by adding "technologies currently under development."[18]

Greening primary steel production is often held out as something only hydrogen can do. About 60% of steel in the United States and Europe is "primary," made from iron ore in a blast furnace with coal-produced coke, and 40% is secondary and hence already an electrified process, made from scrap or recycled steel with an electric arc furnace (EAF). Globally, closer to 80% of steel is primary. Hydrogen can be used for direct reduction of iron (DRI), where it is used to remove oxygen from iron ore so an EAF can convert it to steel, eliminating the need for the coke and blast furnace.

Significantly, "it is important to note that hydrogen in DRI is used for its chemical properties as a reducing agent, and not as a fuel," chemical engineer and hydrogen expert Paul Martin explained in August 2024 in a post on the Hydrogen Science Coalition website.[19] So this solution makes use of hydrogen's value as a chemical feedstock. Hydrogen remains a lousy energy carrier.

On the other hand, achieving the 2% or so emission reductions possible with this substitution in the next quarter century would require a massive and expensive industry retooling—"many hundreds of billions of dollars."[20] It would also require the hydrogen to be truly green with near-zero leakage, which requires another massive investment.

At the same time, a number of companies are pursuing technologies for electrolytic reduction of iron, which would electrify primary steel production and avoid the need for hydrogen generation. We must increase funding for research, development, and demonstration of these technologies to maximize the chance they will be available in the 2030s.

Using hydrogen in primary steel production would have a very high cost per ton of CO_2 saved—much higher than the cost of fully electrifying sectors such as ground transportation and heat for buildings and industry. The entire hydrogen path could become a stranded investment if electrolytic reduction can be commercialized. That's why a massive switch to hydrogen-based steel production does not seem like a good investment for the foreseeable future. Similarly, direct reduction of steel with natural gas is possible, but this does not seem like the optimal near-term strategy until leaks in the gas production and delivery infrastructure are reduced by a factor of ten.

The best near-term strategy for steel is to focus on increasing the recycling rate of scrap steel and hence the production of secondary steel with an EAF. As one 2024 study concluded, "Of all the breakthrough technology options, the scrap-based EAF route is by far the most energy-efficient route, requiring five to seven times less energy than primary steelmaking."[21] As the grid decarbonizes, this would help decarbonize steel. Studies suggest that a strong push could significantly increase steel recycling.[22]

It's clear hydrogen will have limited applications in the future and virtually no place as an energy carrier. Another highly touted use for hydrogen is in the power sector, either for dedicated power generation or helping address the variable power from solar and wind farms. But the round-trip efficiency is terrible, and electrolyzers and fuel cells remain expensive. So Australia's former chief scientist, Alan Finkel, the "architect of the first national clean hydrogen strategy," acknowledged in July 2024 that it was "increasingly clear that green hydrogen is too expensive," and the country was "unlikely to use hydrogen for storage of electricity."[23]

A July 2024 US study on hydrogen found very limited potential for using hydrogen in a carbon-free electric grid.[24] It concluded that "dedicated production of clean hydrogen via electrolysis for use as a power plant fuel is unlikely to yield an effective decarbonization outcome." The study found that even if you could find considerable "free" surplus carbon-free power—wind and solar power (and even nuclear) generated during times of overcapacity—making hydrogen is not very plausible, and "alternative strategies that minimize the need for long-duration storage, such as deploying clean firm generation like geothermal or nuclear, would likely be more cost-effective."[25]

Long-distance shipping and aviation are the two other major economic sectors often identified as good candidates for hydrogen decarbonization. However, it has become increasingly obvious that liquefied green hydrogen is not appropriate for either. Aside from the cost, safety, and infrastructure problems, the overall system efficiency, coupled with the high-GHG leakage and venting rates, negate any climate benefit.

In the case of aviation, a May 2024 BloombergNEF analysis reported that e-kerosene has five to nine times the cost of normal jet fuel. They also project, "By 2050, e-kerosene is likely to still come at a premium compared to conventional jet fuel combined with direct air capture to offset emissions."[26] So e-fuel for jets does not seem like a promising strategy for the foreseeable future. And as noted above, e-ammonia is not a low-GHG fuel for shipping, aside from its high cost, safety problems, and the potential increase in deaths by hundreds of thousands per year. We've also seen that biogenic CO_2 is not a scalable climate solution, whereas if direct air capture is the source of the CO_2, "total cost of ownership for methanol-fueled ships would be almost triple that of conventional vessels."[27]

The best strategy for both the aviation and shipping sectors is to electrify as much of them as possible. Planes for shorter hops can be fully electrified, and longer-distance jets can use batteries in a hybrid configuration to increase fuel economy. Similarly, many types of shipping can be electrified. In 2023, for instance, the Chinese launched the first all-electric cargo ship, capable of carrying up to 700 containers that will "sail a route stretching more than 600 miles" along the Yangtze River to the sea. The ship will "will swap batteries along the route with the batteries recharged at stations along the route."[28]

Yes some subsectors of the economy will be challenging to decarbonize. But the overwhelming majority can make significant strides with efficiency and electrification. It would be a mistake to spend hundreds of billions of dollars to address problems such as intercontinental shipping or aviation when these will not be cost-effective contributors any time soon to the urgent need to rapidly reduce GHG emissions by over 90% by 2050. Also, premature bets could easily result in enormous stranded costs.

In every sector needing a zero-carbon energy carrier, we must pour far more money into all electrification technologies than we do into hydrogen energy technologies because, again, the hydrogen solution to a climate problem is exceedingly unlikely to compete with the electric one. Hydrogen funding should focus on figuring out how to generate genuinely green hydrogen with near-zero leakage for use as a chemical feedstock.

Beyond that, hydrogen energy technologies are unlikely to contribute more than another 1% of overall net GHG emission reductions by 2050. That reality leads to another major conclusion: Neither government policy nor business investment should be based on the belief that hydrogen energy technologies will inevitably be a significant contributor to decarbonizing the energy system or even have meaningful commercial success in the near or medium term.

In other words, hydrogen investments do not deserve a privileged position among the list of potential climate solutions. Quite the reverse. Nothing about hydrogen justifies hundreds of billions of dollars in tax incentives or subsidies over the next fifteen years in the United States or globally. Whatever modest contribution hydrogen energy technologies might make to energy-related GHG emission reductions, investments can and should be slow. We must minimize both the opportunity cost of misallocating vast amounts of renewables and money and the likelihood of massive stranded assets built for a hydrogen economy that never materializes.

Hydrogen is on a path to becoming the biggest boondoggle in the history of climate and energy, undermining the fight to avoid catastrophic and irreversible climate change because of its huge inefficiency, opportunity cost, and leakages.

The 2024 election results underscore the importance of these conclusions. As I have argued, our decades of dawdling mean we must not be distracted by false promises any longer. We must focus our money and effort on rapidly deploying technologies that are already successful—real solutions, such as renewable energy, advanced batteries, and electrification.

Now there will be more dawdling—and even worse, an active effort to undermine the progress we have made both domestically and internationally, which will last until at least 2029 and possibly much longer. "There's no question that the Trump Administration will take aim at every facet of the federal government's climate efforts," as author and climate expert Bill McKibben wrote in a post-election *New Yorker* newsletter. The big question he notes is whether Trump will gut the Inflation Reduction Act. "Expect an effort to protect the most fossil-fuel-friendly provisions," like the huge tax credits for CCS, "and to strip out support for electric vehicles, heat pumps, and solar panels."

Similarly, Trump is expected to weaken standards for green hydrogen and promote hydrogen made from fossil fuels, including dirty blue hydrogen. That would make hydrogen even more of a climate problem than it already is.

Trump is also expected to pursue policies that will increase methane emissions and leaks, such as ending the moratorium on liquefied natural gas export permits and undoing new limits on flaring and leaks from oil and gas operations.

Methane matters because it has a much higher global warming potential over 20 years than CO_2 but a much shorter atmospheric lifetime. Hydrogen also drives near-term warming—in large part because its chemical reactions increase methane's lifetime and atmospheric abundance. So methane and hydrogen are a dangerous duo, and a world with higher usage and emissions of both will see faster climate change, not slower.

On October 24, 2024, the UN Environment Programme released its annual report, *Emissions Gap Report 2024*. The UN warned that the world has a "massive gap between rhetoric and reality" on global climate action, and "continuation of current policies will lead to a catastrophic temperature rise of up to 3.1°C." Donald Trump's victory is but the latest warning that we must ignore the siren song—the hype—of the many false promises described in this book.

Acknowledgments

I must start by expressing my extreme gratitude to sustainability expert Niall Enright, without whom this book might not exist. Enright's unexpected and generous review of the original book on LinkedIn in February 2024 inspired me to revise and update it.

While this book's contents and final judgments are my responsibility alone, I am very grateful to everyone who shared their ideas with me for the 2004 edition and this one.

This revised and updated book builds on the analysis of the core flaws of a hydrogen economy I identified in the original book, thanks to many people who shared their ideas with me two decades ago. So I would like to thank all of them: John Atcheson, Andrew Browning, Stan Bull, Rodney Carlisle, Brian Castelli, Sean Casten, Tom Casten, Steve Chalk, Don Chen, Prashant Chintawar, Helena Chum, Mark Chupka, Jon Coifman, Gregg Cooke, Bob Corell, William Cratty, Peter Dalpe, Kenneth Deffeyes, Raymond Drnevich, Alex Farrell, Stephen Folga, David Garman, Jerry Gillette, Sig Gronich, Hank Habicht, David Hamilton, David Hawkins, Steve Hester, Dale Heydlauff, Dennis Hughes, Tina

Kaarsberg, Greg Kats, George Kehler, David Keith, Jon Koomey, Tom Kreutz, Stephen Kukucha, John Lembo, Mark Levine, James Lide, Michael Love, Andrew Lundquist, Lee Lynd, James McElroy, Alden Meyer, Marianne Mintz, Fred Mitlitsky, Peter Molinaro, William Morgan, Pete O'Connor, Joan Ogden, Jim Olson, Michael Oppenheimer, Peter Pintar, David Redstone, Jennifer Rogers, Jim Rogers, Arthur Rosenfeld, Neil Rossmeissl, George Rudins, Barney Rush, Roger Saillant, Robert San Martin, Jennifer Schafer, Jeff Serfass, Phil Sharp, Polly Shaw, Dale Simbeck, Andrew Skok, David Scott Smith, K. R. Sridhar, Neil Stetson, Peter Teagan, George Thomas, Sandy Thomas, Mark Williams, John Wilson, and William Wilson.

For the 2024 book, I'd like to thank those who shared their ideas with me as I researched the core chapters over the past two years: Danny Cullenward, Howard Herzog, David Ho, Andrew Jones, David Keith, Tim Searchinger, June Sekera, Robert Socolow, Daniel Stephens, John Sterman, and Martin Tengler.

Special thanks go to those who reviewed all or part of the book: Sara Barczak, Ira Arlook, Michael Barnard, Simon Bennett, Scott Denman, Auden Schendler, Marcus Schneider, and Kelly Trout.

I'm grateful to Saul Griffith for providing the Foreword. Over the years, I have learned much from his genius, especially in the area of electrification.

I want to thank Climate Imperative for supporting my work as I researched and wrote this book.

I am very grateful to Michael Mann and Heather Kostick of the University of Pennsylvania's Center for Science, Sustainability, and the Media for their support during the extended research and writing of both this book and its supporting analyses.

I'm deeply appreciative of Rebecca Bright, my editor at Island Press. She supported the idea of this new edition from the beginning, provided invaluable editorial advice, and made the entire process through the final product a very positive experience. I would also like to thank Sharis Simonian and the production staff at Island Press for their hard work.

I am indebted to my late mother for applying her world-class language and editing skills to innumerable drafts of the original book. I miss her every day.

Finally, words cannot describe my gratitude for the gift of my daughter, Antonia, who provides me with unlimited inspiration. Ultimately, it is the Antonias around the world whom we must all bear in mind when we contemplate the consequences of our action—and our inaction—on the climate.

Notes

Introduction

1. International Energy Agency, "Hydrogen," accessed August 2024.
2. John Bockris, "The Origin of Ideas on a Hydrogen Economy and Its Solution to the Decay of the Environment," *International Journal of Hydrogen Energy*, July–August 2002.
3. S&P Global Commodities, "S&P Global Commodity Insights Releases Its Latest 2024 Energy Outlook," December 14, 2023.
4. Alexander Budzier et al., "Five Strategies for Optimizing Power-to-X Projects," Boston Consulting Group, September 28, 2023.
5. Xiaoting Wang, "Electrolysis System Cost Forecast 2050: Higher for Longer," BloombergNEF, September 12, 2024. See also Martin Tengler, head of hydrogen research at BloombergNEF, LinkedIn post, September 2024.
6. International Energy Agency, "Global Hydrogen Review 2023," September 2023.
7. Eric Wesoff, "A Profitable Fuel-Cell Company Finally Emerges amid Industry-wide Losses," June 20, 2023.
8. Ghassan Wakim, "Hydrogen in the Power Sector: Limited Prospects in a Decarbonized Electric Grid," Clean Air Task Force, July 4, 2024.
9. Prachi Patel, "Home Heating with Hydrogen: Ill-Advised as It Sounds, Several Studies Reveal Serious Drawbacks," *IEEE Spectrum*, October 4, 2022. See also Jan Rosenow, "A Meta-Review of 54 Studies on Hydrogen Heating," *Cell Reports Sustainability*, January 26, 2024.

10. Michael Liebreich, "Emission Reduction Using 1Wh of Renewable Energy," chart posted on X.com, October 5, 2023.

11. Fuyuan Yang, "Review on Hydrogen Safety Issues: Incident Statistics, Hydrogen Diffusion, and Detonation Process," *International Journal of Hydrogen Energy Review*, September 3, 2021.

12. International Atomic Energy Agency, "Mitigation of Hydrogen Hazards in Severe Accidents in Nuclear Power Plants," 2011.

13. UN Intergovernmental Panel on Climate Change, "An IPCC Special Report: Global Warming of 1.5°C," 2018.

14. Lori S. Siegel, Andrew P. Jones, John Sterman, et al., En-ROADS update, Climate Interactive, October 2024. See also Lori S. Siegel et al., En-ROADS release notes, October 2024.

15. Maria Sand et al., "A Multi-Model Assessment of the Global Warming Potential of Hydrogen," *Communications Earth & Environment*, June 7, 2023.

16. Adam R. Brandt, "Greenhouse Gas Intensity of Natural Hydrogen Produced from Subsurface Geologic Accumulations," *Joule*, August 2023.

17. The 22% figure comes from Russell Moy, "Tort Law Considerations for the Hydrogen Economy," *Energy Law Journal*, November 2003. The 40% figure comes from James Hansel, "Safety Considerations for Handling Hydrogen," presentation to the Ford Motor Company, June 12, 1998.

18. Max Bearak, "Inside the Global Race to Turn Water into Fuel," *New York Times*, March 11, 2023.

19. Saul Griffith and Dan Cass Energy in conversation with Ebony Bennett, "The Big Switch with Saul Griffith," February 24, 2022.

20. Jeff St. John, "The Case Against the US Government's Big 'Blue Hydrogen' Bet," Canary Media, October 18, 2023.

21. David Schlissel et al., "Blue Hydrogen: Not Clean, Not Low Carbon, Not a Solution," Institute for Energy Economics and Financial Analysis, September 2023.

22. Kobad Bhavnagri, "Blue Hydrogen Could Become the White Elephant on Your Balance Sheet," Bloomberg, December 15, 2021.

23. M. V. Ramana, "The Forgotten History of Small Nuclear Reactors," *IEEE Spectrum*, April 27, 2015.

24. Diana Kinch, "DRI Steel Decarbonization: Direct Reduced Iron and Its Role in Sustainable Steelmaking, *S&P Global*, June 22, 2022.

Chapter 1

1. As quoted in National Research Council, "Energy Research at DOE: Was It Worth It?, Energy Efficiency and Fossil Energy Research 1978 to 2000," 2001.

2. National Academies of Sciences, "The Hydrogen Economy: Opportunities, Costs, Barriers, and R&D Needs," 2004.
3. American Physical Society, "The Hydrogen Initiative," March 2004.
4. Richard Truett, "Volume Fuel Cell Cars at Least 25 Years Away, Toyota Says," *Automotive News*, January 10, 2005.
5. Phil Wrigglesworth, "The Car of the Perpetual Future," *The Economist*, September 6, 2008.
6. Eric Wesoff, "Fuel-Cell Market Update, Greentech Media, April 10, 2017.
7. Ford Motor Company, "Sustainability Report 2013/14," June 18, 2014.
8. David Waterworth, "Toyota to Lose $100,000 on Every Hydrogen FCV Sold," CleanTechnica, November 19, 2014.
9. US Department of Energy, "History of the Electric Car," September 15, 2014.
10. Rachel Morison and Petra Sorge, "Europe's Spending Billions on Green Hydrogen. It's a Risky Gamble," Bloomberg, May 17, 2024.
11. Ford Motor Company, "2024 Integrated Sustainability and Financial Report," 2024.
12. Forvia Hella, "Air Liquide and Faurecia Announce Development Agreement to Boost Hydrogen for Heavy-Duty Vehicles," Air Liquide, October 12, 2021.
13. Eric Schaub, "Home Heating with Hydrogen Is Ill-Advised," *IEEE Spectrum*, October 4, 2022.
14. "After Many False Starts, Hydrogen Power Might Now Bear Fruit," *The Economist*, July 4, 2020.
15. Leigh Collins, "Hydrogen Vehicle Registrations Fell by More Than 50% in South Korea in 2023, Government Figures Reveal," *Hydrogen Insight*, January 13, 2024.
16. Rachel Morison and Petra Sorge, "Europe's Spending Billions on Green Hydrogen. It's a Risky Gamble," Bloomberg, May 17, 2024.
17. Electric Power Research Institute, "Impacts of IRA's 45V Clean Hydrogen Production Tax Credit," November 3, 2023.
18. Jeff St. John, "Clean Energy Experts Break Down Hydrogen Hype and Hope," Canary Media, February 1, 2024. In this video interview, hydrogen expert Michael Liebreich says, "It's not a complete boondoggle it is an almost complete boondoggle." But three minutes later, after explaining that the market is unlikely to finance and build most hydrogen projects, he said, "Maybe I've even been unfair calling it an almost complete boondoggle. It's an almost complete distraction."

Chapter 2

1. John Rigden, *Hydrogen: The Essential Element*, Harvard University Press, 2002. Rigden's full quote is, "With one exception—the helium atom—hydrogen is the mother of all atoms and molecules."

2. Jules Verne, *The Mysterious Island*, 1875.
3. Max Pemberton, The Iron Pirate, Greycoine Book Manufacturing Company, 1894.
4. Jeremy Bernstein, "Recreating the Power of the Sun," *New York Times Magazine*, January 3, 1982.
5. Lawrence Lidsky, "The Trouble with Fusion," *Technology Review*, October 1983.
6. Lawrence Livermore National Laboratory, "Frequently Asked Questions," accessed August 2024.
7. Eric Schaub, "National Ignition Facility: Impractical," *IEEE Spectrum*, August 5, 2024. The quote: "In 2014, the facility heralded a fusion breakthrough called fuel gain, an important step toward ignition. However, it turned out that NIF's researchers had made the assessment using only the laser energy that hit the capsule, discounting the vast majority of the lasers' energy in the process."
8. Geoff Brumfiel, "EU Research Funds to Be Diverted to Fusion Reactor," *Nature*, July 7, 2010. See also Daniel Clery, "Fusion Megaproject Confirms 5-Year Delay, Trims Costs," *Science*, June 16, 2016.
9. Katharine Hayhoe, "Science Festival Envisions Our Planet's Future," *LinkedIn Pulse*, May 21, 2024.
10. Aproop Dheeraj Ponnada, "Analyzing Climate 'Silver Bullets' with System Dynamics," *Climate Interactive*, August 29, 2024.

Chapter 3

1. Grace Snelling, "Green, Blue, Gold, and More: What the Different Colors of Hydrogen Mean," *Fast Company*, December 4, 2023.
2. US Department of Energy, "National Clean Energy Strategy and Roadmap," June 2023.
3. International Energy Agency, "The Future of Hydrogen," June 2019.
4. US Department of Energy, "Hydrogen Resources," Office of Energy Efficiency and Renewable Energy, 2024.
5. Kamala Schelling, "Green Hydrogen to Undercut Gray Sibling by End of Decade," BloombergNEF, 2023.
6. US Department of Energy, Office of Fossil Energy, "FutureGen—A Sequestration and Hydrogen Research Initiative," fact sheet, February 2003.
7. Dale Simbeck, "Long-Term Technology Pathways to Stabilization of Greenhouse Gas Concentrations," presentation to the Aspen Global Change Institute, Aspen, Colorado, July 6–13, 2003.
8. Matthew Wald, "Coal Industry Finds a New Way to Cut Emissions," *New York Times*, December 18, 2007.
9. Peter Folger, "The FutureGen Carbon Capture and Sequestration Project: A

Brief History and Issues for Congress," Congressional Research Service, February 10, 2014.

10. Global Energy Monitor, "Retrofitting the Existing Coal Fleet with Carbon Capture Technology," US Department of Energy, December 2008.

11. World Nuclear Association, "Hydrogen Production and Uses," May 1, 2024.

12. Kamala Schelling, "Green Hydrogen to Undercut Gray Sibling by End of Decade," BloombergNEF, December 12, 2023.

13. Rachel Morison and Petra Sorge, "Europe's Spending Billions on Green Hydrogen. It's a Risky Gamble," Bloomberg, May 17, 2024.

14. Rachel Parkes, "Cost of Electrolysers for Green Hydrogen Production Is Rising Instead of Falling: BNEF," *Hydrogen Insight*, March 2024.

15. S&P Global, "S&P Global Commodity Insights Releases Its Latest 2024 Energy Outlook," *S&P Global*, December 14, 2023.

16. Alexander Budzier et al., "Five Strategies for Optimizing Power-to-X Projects," Boston Consulting Group, September 28, 2023.

17. Martin Tengler, head of hydrogen research at BloombergNEF, LinkedIn post, September 2024; see also Xiaoting Wang, "Electrolysis System Cost Forecast 2050: Higher for Longer," BloombergNEF, September 12, 2024.

18. Abhishek Malhotra and Tobias S. Schmidt, "Accelerating Low-Carbon Innovation," *Joule*, November 2020.

19. Wilson Ricks et al., "Minimizing Emissions from Grid-Based Hydrogen Production in the United States," *Environmental Research Letters*, January 6, 2023.

20. Colton Poore, "Without Guidance, Inflation Reduction Act Tax Credit May Do More Harm Than Good," *Princeton News*, December 20, 2022.

21. Wilson Ricks et al., "Minimizing Emissions from Grid-Based Hydrogen Production in the United States," *Environmental Research Letters*, January 6, 2023.

22. Mark Trexler, *Are Carbon Offsets Helping Tackle Climate Change?* Udemy online course, November 2022.

23. Anders Bjørn et al., "Renewable Energy Certificates Threaten the Integrity of Corporate Science-Based Targets," *Nature Climate Change*, June 2022.

24. Ben Elgin and Sinduja Rangarajan, "What Really Happens When Emissions Vanish," Bloomberg, October 31, 2022.

25. Jeff St. John, "Clean Energy Experts Break Down Hydrogen Hype and Hope," Canary Media, February 1, 2024.

26. Eric Schaub, "Home Heating with Hydrogen Is Ill-Advised," *IEEE Spectrum*, October 1, 2022.

27. Michael Liebreich, "Emission Reduction Using 1Wh of Renewable Energy," chart posted on X.com, October 5, 2023.

28. US Department of Energy, "Gasification Workshop Presentation," Office of Energy Efficiency and Renewable Energy, November 2022.

29. Jennifer Pett-Ridge et al., "Roads to Removal: Options for Carbon Dioxide Removal in the United States," Lawrence Livermore National Laboratory, December 2023.
30. Sam Meredith, "Natural Hydrogen: A New Gold Rush for a Potential Clean Energy Source," CNBC, March 26, 2024.
31. Arnout Everts, "Everything You Need to Know About Natural or Geologic Hydrogen," H2 Science Coalition, March 14, 2024.
32. Clive Cookson, "Geologists Signal Start of Hydrogen Energy 'Gold Rush,'" *Financial Times,* February 18, 2024.
33. Fred Pearce, "Natural Geologic Hydrogen: A New Frontier in Climate Change Mitigation," *Yale E360*, January 25, 2024.
34. Rystad Energy, "White Gold Rush and the Pursuit of Natural Hydrogen," March 13, 2024.
35. Eric Wesoff, "The $245M Bid to Pull Clean Hydrogen Straight from the Earth," Canary Media, February 14, 2024.
36. Tim McDonnell, "Why Investors Are Betting Big on Geologic Hydrogen," *Semafor*, February 16, 2024.
37. William Allstetter, "Colorado's Hydrogen Geological Hunt: The Investment Gold Rush," *The Colorado Sun*, April 1, 2024.
38. Sam Meredith, "Natural Hydrogen: A New Gold Rush for a Potential Clean Energy Source," CNBC, March 26, 2024.
39. International Energy Agency, "Global Hydrogen Review 2023," September 2023.
40. Xia Ri, "Commercial Production of White Hydrogen Unlikely in the Short Term," *Anbound*, March 4, 2024.
41. Jennifer Couzin-Frankel, "Breakthrough of the Year 2023," *Science*, December 22, 2023.
42. Steven R. Helfrich and John T. Robinson, "A Deep Reservoir for Hydrogen Drives Intense Degassing in the Bulqizë Ophiolite," *Science*, February 8, 2024.
43. Eric Hand, "Gusher Gas: Deep Mine Stokes Interest in Natural Hydrogen," *Science*, February 8, 2024.
44. Adam R Brandt, "Greenhouse Gas Intensity of Natural Hydrogen Produced from Subsurface Geologic Accumulations," *Joule*, August 16, 2023.
45. Andrews R. George, "Exploding Mine: Albania's Bulqize and Its Hydrogen Potential," *National Geographic*, March 7, 2024.
46. Marta Zaraska, "Once-Hidden Hydrogen Gas Deposits Could Be a Boon for Clean Energy," *Scientific American*, May 8, 2024.
47. "Fever," *Cambridge Dictionary*, accessed August 2024.
48. Michael Le Page, "Is Geological Hydrogen a Green Solution to Our Energy Needs?," *New Scientist*, July 27, 2023.

49. Maria Sand et al., "A Multi-Model Assessment of the Global Warming Potential of Hydrogen," *Communications Earth & Environment*, June 7, 2023. See also Alissa B. Ocko and Steven P. Hamburg, "Climate Consequences of Hydrogen Emissions," *Atmospheric Chemistry and Physics*, July 20, 2022; Didier Hauglustaine et al., "Climate Benefit of a Future Hydrogen Economy," *Nature Communications*, November 26, 2022.

50. Arnout Everts, "Everything You Need to Know About Natural or Geologic Hydrogen," *H2 Science Coalition*, March 14, 2024.

51. Joseph Romm, "Are Carbon Offsets Unscalable, Unjust, and Unfixable—and a Threat to the Paris Climate Agreement?," Center for Science, Sustainability, and the Media, University of Pennsylvania, June 27, 2023.

Chapter 4

1. Hannah Ritchie and Max Roser, "Nuclear Energy," *Our World in Data*, July 2020, revised in April 2024.

2. Alexander Budzie, "Five Strategies for Optimizing Power-to-X Projects," September 28, 2023.

3. Patty Durand, "Plant Vogtle: Not a Star, but a Tragedy for the People of Georgia," *Power Magazine*, August 11, 2023.

4. Bill Gates, *Face the Nation*, June 16, 2024. Asked what the Wyoming plant will cost, Gates, replied, "Well, if you count all the first of a kind costs, you know, where we've been working for many years designing this thing, you could get a number close to 10 billion."

5. David Greenberg, "The Forgotten History of Small Nuclear Reactors," *IEEE Spectrum*, April 27, 2015.

6. J. C. Miller and K. J. S. Schaefer, "Hydrogen Production and Uses," *ScienceDirect*, April 2024.

7. US Department of Energy, "3 Nuclear Power Plants Gearing Up for Clean Hydrogen Production," November 9, 2022.

8. Reuters, "Nuclear Gets a Boost from Europe's New Green Energy Targets," September 13, 2023.

9. Westinghouse. "Westinghouse Explores Clean Hydrogen Production at Nuclear Power Plants," September 20, 2023.

10. Reuters, "Nuclear-Hydrogen Marriage Has Potential, U.S. Energy Loans Chief Says," September 28, 2023.

11. Sonal Patel, "Constellation Planning Significant Nuclear-Powered Hydrogen Facility at LaSalle," October 19, 2023.

12. Uma Gupta, "NPCIL to Pilot Nuclear-Powered Hydrogen Generation," *PV Magazine*, November 20, 2023.

13. Kazunari Hanawa and Ryosuke Matsuzoe, "Japan Eyes Hydrogen Production Using Next-Gen Nuclear Reactor," *Nikkei Asia*, April 4, 2024.
14. NucNet, "OKG to Supply Hynion with Clean Hydrogen from Oskarshamn Nuclear Station," May 5, 2024.
15. Leigh Collins, "South Korea's First Nuclear Hydrogen Project to Be Built by 2027 by Eight Partners Including Hyundai and Samsung," *Hydrogen Insight*, June 19, 2024.
16. Jim Polson, "Half of America's Nuclear Power Plants Seen as Money Losers," Bloomberg, June 14, 2017.
17. Michel Berthélemy and Lina Escobar Rangel, "Nuclear Reactors' Construction Costs: The Role of Lead-Time, Standardization and Technological Progress," *Energy Policy*, July 2015.
18. Rachel Parkes, "Installing Clean Hydrogen Production in Existing Nuclear Power Plants Would Be a Permitting Nightmare: Report," *Hydrogen Insight*, December 2023.
19. Ghassan Wakim and Kasparas Spokas, "Hydrogen in the Power Sector: Limited Prospects in a Decarbonized Electric Grid," Clean Air Task Force, June 2024.
20. Thomas Wellock, "'Too Cheap to Meter': A History of the Phrase," Nuclear Regulatory Commission, updated September 24, 2021.
21. US Nuclear Regulatory Commission, "Report on the Accident at Three Mile Island," October 1979.
22. Arnulf Grubler, "The Costs of the French Nuclear Scale-Up: A Case of Negative Learning by Doing," *Energy Policy*, September 2010.
23. Michael Cembalest, "Eye on the Market," JP Morgan Asset and Wealth Management, April 2, 2024.
24. Massachusetts Institute of Technology, *The Future of Nuclear Power: An Interdisciplinary MIT Study*, 2003.
25. Joseph Romm, Statement before the Senate Committee on Environment and Public Works Subcommittee on Clean Air and Nuclear Safety, July 16, 2008.
26. Reuters, "U.S. Power Plant Costs Up 130% Since 2000—CERA," February 14, 2008.
27. Nuclear Engineering International, "How Much? For Some Utilities, the Capital Costs of a New Nuclear Power Plant Are Prohibitive," November 20, 2007.
28. *New York Times*, "Atomic Balm?," July 16, 2006.
29. Rebecca Smith, "New Wave of Nuclear Plants Faces High Costs," *Wall Street Journal*, June 14, 2008.
30. Jeff Amy, "US Energy Secretary Calls for More Nuclear Power While Celebrating $35 Billion Georgia Reactors," Associated Press, June 2, 2024.

31. Mycle Schneider et al., "The World Nuclear Industry Status Report 2022," World Nuclear Report, November 18, 2022.

32. Michael Cembalest, "14th Annual Energy Paper," JP Morgan Asset and Wealth Management, 2024.

33. UK Government, "Department for Business, Enterprise & Regulatory Reform. Nuclear Power: A Review of the Potential," 2009.

34. Simon Jack, "Hinkley C: UK Nuclear Plant Price Tag Could Rocket by a Third," BBC News, January 23, 2024.

35. Francois De Beaupuy, "Hinkley Point Nuclear Plant in UK Stops Getting Funding from China's CGN," Bloomberg, December 13, 2023.

36. Jim Green, "Small Modular Nuclear Reactors: A History of Failure," Climate & Capital Media, January 17, 2024.

37. M. V. Ramana and Zia Mian, "One Size Doesn't Fit All: Social Priorities and Technical Conflicts for Small Modular Reactors," *Energy Research & Social Science*, June 2014.

38. Rod Ewing, "Small Modular Reactors Produce High Levels of Nuclear Waste," *Stanford Report*, May 30, 2022.

39. Lindsay Krall et al., "Nuclear Waste from Small Modular Reactors," *Proceedings of the National Academy of Sciences*, May 31, 2022.

40. M. V. Ramana and Zia Mian, "One Size Doesn't Fit All: Social Priorities and Technical Conflicts for Small Modular Reactors," *Energy Research & Social Science*, June 2014.

41. Richard Vietor, "NuScale: Analyzing the Potential of Small Modular Reactors," *Harvard Business School*, June 2016.

42. Caroline Delbert, "This Tiny Nuclear Reactor Will Change Energy—and Now It's Officially Safe," *Popular Mechanics*, September 2, 2020.

43. Will Wade, "First U.S. Small Nuclear Project Canceled After Costs Climb 53%," *Financial Post*, November 8, 2023.

44. Ed Lyman, "Five Things the Nuclear Bros Don't Want You to Know About Small Modular Reactors," Union of Concerned Scientists, April 30, 2024.

45. Casey Crownhart, "We Were Promised Smaller Nuclear Reactors. Where Are They?," *Technology Review*, February 8, 2023.

46. Mycle Schneider et al., "The World Nuclear Industry Status Report 2022," World Nuclear Report, November 18, 2022.

47. David Greenberg, "The Forgotten History of Small Nuclear Reactors," *IEEE Spectrum*, April 27, 2015.

48. Fanny Böse et al., "Questioning Nuclear Scale-Up Propositions: Availability and Economic Prospects of Light Water, Small Modular and Advanced Reactor Technologies," *Energy Research and Social Science*, April 2024.

49. Nuclear Energy Institute, "Nuclear Costs in Context," October 2018.

50. Benjamin Sovacool and M. V. Ramana, "Back to the Future: Small Modular Reactors, Nuclear Fantasies, and Symbolic Convergence," *Science, Technology, & Human Values*, July 23, 2014.

51. Dr. Edwin Lyman, "Small Nuclear Reactor Contract Fails, Signaling Larger Issues in Nuclear Energy Development," Union of Concerned Scientists, November 9, 2023.

52. Shanlai Lu, "Evaluation of NuScale Post ECCS Actuation Boron Dilution Events," National Academies Press, July 6, 2020.

53. Arjun Makhijani and M. V. Ramana, "Questions for NuScale VOYGR Reactor Certification: When Will It Be Done? And Then, Will It Be Safe?," Environmental Working Group, April 9, 2023.

54. Idaho National Laboratory, "Pilot Projects Test a Path to Hydrogen Economy," October 8, 2020.

55. Clean Air Task Force, "Solid Oxide Electrolysis Technology Status Assessment," November 15, 2023.

56. John Kemeny et al., "Report of the Commission on the Accident at Three Mile Island—The Need for Change: The Legacy of TMI," October 1979.

57. International Atomic Energy Agency, *Mitigation of Hydrogen Hazards in Severe Accidents in Nuclear Power Plants*, 2011.

58. Robert Ring et al., "Modeling the Risk of Fire/Explosion Due to Oxidizer/Fuel Leaks in the Ares I Interstage," NASA Technical Reports, October 21, 2008.

59. Gniewomir Flis and Ghassan Wakim, "Solid Oxide Electrolysis Technology Status Assessment," Clean Air Task Force, November 2023.

Chapter 5

1. Jeff St. John, "The Case Against the U.S. Government's Big Blue Hydrogen Bet," Canary Media, 2023.

2. Polly Martin, "Why Making Hydrogen from Coal Could Be Better for the Planet than Blue H_2 Derived from Natural Gas," *Hydrogen Insight*, December 7, 2023.

3. Anika Juhn, "Blue Hydrogen: Not Clean, Not Low Carbon, Not a Solution" Institute for Energy Economics and Financial Analysis, September 2023.

4. Kobad Bhavnagri, "Market Risks: White Elephant in Push for Blue Hydrogen," Bloomberg, December 28, 2021. Bloomberg says of blue hydrogen that "the technology only works at large scale."

5. International Renewable Energy Agency, "Geopolitics of the Energy Transformation: The Hydrogen Factor," 2022.

6. Deborah Gordon et al., "The Evaluating Net Life-Cycle Greenhouse Gas Emissions Intensities from Gas and Coal at Varying Methane Leakage Rates," *Environmental Research Letters*, July 17, 2023.

7. Robert W. Howarth and Mark Z. Jacobson, "How Green Is Blue Hydrogen?," *Energy Science & Engineering*, August 12, 2021.

8. Robert W. Howarth, "The Greenhouse Gas Footprint of Liquefied Natural Gas (LNG) Exported from the United States," *Energy Science & Engineering*, October 2024. The study concludes that "the greenhouse gas footprint for LNG as a fuel source is 33% greater than that for coal when analyzed using GWP20. . . . Even considered on the time frame of 100 years after emission (GWP100), which severely understates the climatic damage of methane, the LNG footprint equals or exceeds that of coal."

9. Selina Roman-White et al., "Life Cycle Greenhouse Gas Perspective on Exporting Liquefied Natural Gas from the United States: 2019 Update," National Energy Technology Laboratory, September 12, 2019.

10. National Energy Technology, "Hydrogen Shot Technology Assessment: Thermal Conversion Approaches," National Energy Technology Laboratory, December 5, 2023.

11. Leigh Collins, "Why Multi-Million Dollar Blue Hydrogen Investments Might Fast End Up as Stranded Assets," *Recharge News*, January 20, 2022.

12. Oliver Morton, tweet, October 14, 2021.

13. Hiroko Tabuchi and Brad Plumer, "Is This the End of New Pipelines?," *New York Times*, July 8, 2020. The articles notes that in 2020, the Atlantic Coast natural gas pipeline was canceled "after environmental lawsuits and delays had increased the estimated price tag of the project to $8 billion from $5 billion."

14. Benoît Morenne and Joe Barrett, "A New Nimbyism Blocks Carbon Pipelines," *Wall Street Journal*, September 30, 2023.

15. Christina Hoicka et al., "Letter from Scientists, Academics, and Energy System Modellers: Prevent Proposed CCUS Investment Tax Credit from Becoming a Fossil Fuel Subsidy," January 19, 2022.

16. IPCC, "Climate Change 2022: Mitigation of Climate Change," 2022.

17. Vaclav Smil, "Energy at the Crossroads," OECD Global Science Forum, May 17–18, 2006. For a similar calculation, see also Niall Mac Dowell et al., "The Role of CO_2 Capture and Utilization in Mitigating Climate Change," *Nature Climate Change*, April 2017.

18. Jeff St. John, "The Case Against the U.S. Government's Big Blue Hydrogen Bet," Canary Media, October 18, 2023.

19. Saudi Aramco, "Saudi Aramco Will Be Among Largest Investors in Blue Hydrogen," *Hydrogen Central*, December 2023.
20. Jeff St. John, "The Case Against the U.S. Government's Big Blue Hydrogen Bet," Canary Media, October 18, 2023.
21. Vatche Kourkejian, "Carbon Capture, Utilisation and Storage in the GCC," Roland Berger, December 4, 2023.
22. Megren Almutairi and Karen E. Young, "How Carbon Capture Technology Could Maintain Gulf States' Oil Legacy," Columbia University, February 19, 2024.
23. Ahmed Abdulla et al., "Explaining Successful and Failed Investments in U.S. Carbon Capture and Storage Using Empirical and Expert Assessments," *Environmental Research Letters*, 2021.
24. Howard Herzog, Edward Rubin, and Gary Rochelle, "Comment on 'Reassessing the Efficiency Penalty from Carbon Capture in Coal-Fired Power Plants,'" *Environmental Science and Technology*, May 12, 2016.
25. Kurt House et al., "The Energy Penalty of Post-Combustion CO_2 Capture & Storage and Its Implications for Retrofitting the U.S. Installed Base," *Energy & Environmental Science*, January 2009.
26. National Petroleum Council, *Meeting the Dual Challenge: A Roadmap to At-Scale Deployment of Carbon Capture, Use, and Storage*, Chapter 5, July 2020.
27. Niall MacDowell et al., "The Role of CO_2 Capture and Utilization in Mitigating Climate Change," *Nature Climate Change*, April 2017.
28. Jasmin Kemper, "IEA Net Zero Roadmap Update," IEA, November 2023.
29. Jamie Condliff, "Clean Coal's Flagship Project Has Failed," *Technology Review*, June 29, 2017.
30. Wang Ucilia, "NRG's 1B Bet to Show How Carbon Capture Could Be Feasible for Coal Power Plants," *Forbes*, July 15, 2014.
31. Karin Rives, "DOE Spent Nearly $700M on Ill-Fated Coal Carbon Capture Projects, GAO Says," *S&P Global Market Intelligence*, December 27, 2021.
32. Simon, "Petra Nova: Leading CCS Power Station to Provide Model for Australia," CleanTechnica, June 12, 2017.
33. David Schlissel, "Boundary Dam 3 Coal Plant Achieves Goal of Capturing 4 Million Metric Tons of CO_2 but Reaches the Goal Two Years Late," Institute for Energy Economics and Financial Analysis (IEEFA), April 2021; David Schlissel, "IEEFA: Carbon Capture Goals Miss the Mark at Boundary Dam 3 Coal Plant," IEEFA, April 20, 2021.
34. Carlos Anchondo, "CCS 'Red Flag?' World's Sole Coal Project Hits Snag," *E&E News*, January 10, 2022. Challenges with the main CO_2 compressor motor forced its CCS system to "go offline for multiple months last year."

35. Adam Morton, "Chevron Could Be Forced to Pay $100m for Failure to Capture Carbon Emissions," *The Guardian*, June 3, 2020; Adam Morton, "'A Shocking Failure': Chevron Criticised for Missing Carbon Capture Target at WA Gas Project," *The Guardian*, July 19, 2021. The project only started operating in August 2019 "due to technical issues."

36. The Climate Council, "What Is Carbon Capture and Storage?," August 2, 2023.

37. Yuting Zhang et al., "An Estimate of the Amount of Geological CO_2 Storage over the Period of 1996–2020," *Environmental Science & Technology Letters*, July 19, 2022.

38. Briana Mordick and George Peridas, "Strengthening the Regulation of Enhanced Oil Recovery to Align It with the Objectives of Geologic Carbon Dioxide Sequestration," Natural Resources Defense Council, November 2017.

39. International Energy Agency (IEA), "Whatever Happened to Enhanced Oil Recovery?," November 2018.

40. IEA, "Storing CO_2 Through Enhanced Oil Recovery," November 2015.

41. June Sekera and Andreas Lichtenberger, "Assessing Carbon Capture: Public Policy, Science, and Societal Need," *Biophysical Economics and Sustainability*, October 2020.

42. Paulina Jaramillo et al., "Life Cycle Inventory of CO_2 in an Enhanced Oil Recovery System," *Environmental Science & Technology*, September 2009.

43. Deepika Nagabhushan and John Thompson, "Carbon Capture & Storage in the United States Power Sector: The Impact of 45Q Federal Tax Credits," Clean Air Task Force, February 2019.

44. Michael Godec, "Establishing a Business Case for CO_2-EOR with Storage," Advanced Resources International, Inc., September 2018.

45. Mella McEwen, "EOR Technology Opens Door to Billions of Barrels in Residual Oil Zones," *Midland Reporter-Telegram*, October 15, 2016.

46. Kelly Trout et al., "Existing Fossil Fuel Extraction Would Warm the World Beyond 1.5°C," *Environmental Research Letters*, May 17, 2022. Kelly Trout, in personal communications, indicated that meeting the Paris target of "well below 2°C" warming would mean shutting down 17% of existing oil fields. See also "Shut Down Fossil Fuel Production Sites Early to Avoid Climate Chaos, Says Study," *Guardian*, May 17, 2022.

47. David Roberts, "Could Squeezing More Oil out of the Ground Help Fight Climate Change?," *Vox*, December 6, 2019.

48. Nicholas Kusnetz, "Exxon Touts Carbon Capture as a Climate Fix, but Uses It to Maximize Profit and Keep Oil Flowing," *Inside Climate News*, September 27, 2020.

49. J. Russell George, Department of Treasury Letter, April 15, 2020.
50. Christina Hoicka et al., "Letter from Scientists, Academics, and Energy System Modellers: Prevent Proposed CCUS Investment Tax Credit from Becoming a Fossil Fuel Subsidy," January 19, 2022. The letter also states, "Despite decades of research, CCUS is neither economically sound nor proven at scale, with a terrible track record and limited potential to deliver significant, cost-effective emissions reductions."
51. Leah Douglas, "U.S. Carbon Pipeline Company Pledges No Oil Recovery, Bakken Drillers Want It," Reuters, March 11, 2024.
52. Summit Carbon Solutions, "Enhanced Oil Recovery," accessed August 2024.
53. Michael Cembalest, "2021 Annual Energy Paper," JP Morgan Asset and Wealth Management, 2021.
54. Benoît Morenne and Joe Barrett, "A New Nimbyism Blocks Carbon Pipelines," *Wall Street Journal*, September 30, 2023.
55. Jeffrey Tomich, "Developer Scuttles Plans for Midwest CO_2 Pipeline," *Greenwire*, October 20, 2023.
56. E. Larson et al., *Net-Zero America: Potential Pathways, Infrastructure, and Impacts*, final report, Princeton University, October 29, 2021.
57. Pavel Tcvetkov, "Public Perception of Carbon Capture and Storage: A State-of-the-Art Overview," *Heliyon*, December 2019.
58. IPCC, "Special Report Global Warming of 1.5 °C," 2021.
59. National Petroleum Council Report, *Meeting the Dual Challenge: A Roadmap to At-Scale Deployment of Carbon Capture, Use, and Storage*, Chapter 6: CO_2 Transport, December 2019.
60. Julia Simon, "The U.S. Is Expanding CO_2 Pipelines. One Poisoned Town Wants You to Know Its Story," NPR, September 25, 2023.
61. Mike Soraghan and Carlos Anchondo, "Are CO_2 Pipelines Safe?," *EnergyWire*, March 31, 2022.
62. Richard Kuprewicz, "Accufacts' Perspectives on the State of Federal Carbon Dioxide Transmission Pipeline Safety Regulations as It Relates to Carbon Capture, Utilization, and Sequestration Within the U.S.," prepared for the Pipeline Safety Trust, March 23, 2022.
63. Dan Zegart, "The Gassing of Satartia," *HuffPost*, August 26, 2021; Erin Jordan, "Federal Rules for CO_2 Pipelines Outdated, 'Could Threaten Safety,'" Watchdog Report," *Iowa Gazette*, April 1, 2022.
64. Emily Pontecorvo, "CO_2 Pipelines Are Coming. A Pipeline Safety Expert Says We're Not Ready," *Grist*, April 18, 2022.
65. IPCC, "Special Report on Carbon Dioxide Capture and Storage," 2005.
66. Catherine Ruprecht and Ronald Falta, "Comparison of Supercritical and

Dissolved CO_2 Injection Schemes," *Proceedings, TOUGH Symposium*, Lawrence Berkeley National Laboratory, September 2012.

67. Bergur Sigfusson et al., "Solving the Carbon-Dioxide Buoyancy Challenge: The Design and Field Testing of a Dissolved CO_2 Injection System," *International Journal of Greenhouse Gas Control*, February 2015.

68. N. P. Qafoku et al., "A Critical Review of the Impacts of Leaking CO_2 Gas and Brine on Groundwater Quality," Pacific Northwest National Laboratory, September 2015.

69. Duke University, "Leaks from CO_2 Stored Deep Underground Could Contaminate Drinking Water," November 8, 2010, and Mark G. Little and Robert B. Jackson, "Potential Impacts of Leakage from Deep CO_2 Geosequestration on Overlying Freshwater Aquifers," *Environmental Science & Technology*, October 26, 2010. See also Mark G. Little and Robert B. Jackson, "Response to Comment on 'Potential Impacts of Leakage from Deep CO_2 Geosequestration on Overlying Freshwater Aquifers,'" *Environmental Science & Technology*, March 7, 2011.

70. GAO, "Drinking Water: EPA Needs to Collect Information and Consistently Conduct Activities to Protect Underground Sources of Drinking Water," March 2016.

71. Briana Mordick and George Peridas, "Strengthening the Regulation of Enhanced Oil Recovery to Align It with the Objectives of Geologic Carbon Dioxide Sequestration," Natural Resources Defense Council, November 2017 (hereafter NRDC 2017).

72. Stephanie Joyce, "What Happened in Midwest? The Mysterious Gas Leak That Shuttered a School," *Wyoming Public Radio*, November 5, 2016.

73. Ian Duncan et al., "Risk Assessment for Future CO_2 Sequestration Projects Based CO_2 Enhanced Oil Recovery in the U.S.," *Energy Procedia*, February 2009.

74. L. Skinner, "CO_2 Blowouts: An Emerging Problem," *World Oil*, January 2003.

75. Sean Porse et al., "Can We Treat CO_2 Well Blowouts Like Routine Plumbing Problems? A Study of the Incidence, Impact, and Perception of Loss of Well Control," *Energy Procedia*, December 2014.

76. A. Lustgarten, "Injection Wells—The Poison Beneath Us," ProPublica, June 21, 2012. See also *NRDC 2017*, which explains, "Mechanical integrity violations can be issued for a number of reasons, from actual mechanical flaws in injection wells to incomplete records or paperwork."

77. Briana Mordick and George Peridas, "Strengthening the Regulation of Enhanced Oil Recovery to Align It with the Objectives of Geologic Carbon Dioxide Sequestration," Natural Resources Defense Council, November 2017. The report explains, "'Enhanced recovery' is the term EPA uses to describe

Class II permits for wells injecting fluids consisting of brine, freshwater, steam, polymers, or carbon dioxide into oil-bearing formations to recover residual oil and, in limited applications, natural gas. These can include wells injecting fluids for either secondary recovery (water flooding) or tertiary recovery (often. referred to as enhanced oil recovery)."

78. Dan Zegart, "The Gassing of Satartia," *HuffPost*, August 26, 2021.

79. "Denbury Pays Big Fine for 2011 Oil Well Blowout," Associated Press, July 25, 2013 as cited in Global Energy Monitor wiki.

80. Sanne Akerboom, "Different This Time? The Prospects of CCS in the Netherlands in the 2020s," *Frontiers in Energy Research*, May 4, 2021.

81. Von Jessica Donath, "One German Town's Fight Against CO_2 Capture Technology," *Spiegel International*, August 20, 2010; Terry Slavin and Alok Jha, "Not Under Our Backyard, Say Germans, in Blow to CO_2 Plans," *The Guardian*, July 29, 2009.

82. Philippa Parmiter and Rebecca Bell Date, "Public Perception of CCS: A Review of Public Engagement for CCS Projects," EU CCUS Projects Network, May 2020.

83. Emma Martin-Roberts et al., "Carbon Capture and Storage at the End of a Lost Decade," *One Earth*, November 19, 2021.

84. IPCC, *SRCCS 2005*, "Chapter 5: Underground Geological Storage."

85. EPA, "Class VI - Wells Used for Geologic Sequestration of Carbon Dioxide," *S&P Global*, May 3, 2022.

86. Brandon Mulder, "Texas Seeks Regulatory Authority from EPA over CO_2 Injection Wells," *S&P Global*, May 3, 2022.

87. Nicholas Kusnetz, "Proponents Say Storing Captured Carbon Underground Is Safe, But States Are Transferring Long-Term Liability for Such Projects to the Public," *Inside Climate News*, April 26, 2022.

88. Scott Anderson and Nichole Saunders, "RE: Council on Environmental Quality Carbon Capture, Utilization, and Sequestration Guidance," Environmental Defense Fund, April 13, 2022. See also Scott Anderson, "States Should Not Weaken Liability Laws for CCS Projects," EDF, May 3, 2022.

89. Statoil, Reply to Carbon Disclosure Project 2013 Information Request, 2013.

90. Richard Monastersky, "Seabed Scars Raise Questions over Carbon Storage Plan," *Nature*, December 2013.

91. Karolin Schaps, "Subsea Ravine Leaks a New Headache for Carbon Capture," Reuters, September 17, 2012.

92. Jonny Rutqvist, "The Geomechanics of CO_2 Storage in Deep Sedimentary Formations," *Geotechnical and Geological Engineering*, January 12, 2012.

93. Joshua A. White et al., "Geomechanical Behavior of the Reservoir and Caprock System at the In Salah CO_2 Storage Project," *PNAS*, May 27, 2014.

94. P. S. Ringrose, "The In Salah CO_2 Storage Project Lessons Learned and Knowledge Transfer," *Energy Procedia*, 2013.

95. InSAR allows millimeter-scale changes in ground surface elevation to be monitored and has proven especially valuable at the In Salah CCS site. By combining satellite data with rock mechanical models, InSAR has been used to monitor the geomechanical response related to subsurface pressure changes caused by CO_2 injection. This technology addresses a key question for CO_2 storage, namely control and management of injection pressure.

Chapter 6

1. David Keith, August 18, 2022, personal communications with the author.

2. Meron Tesfaye, "Priming the E-Pump: Explaining E-Fuels," The Bipartisan Policy Center, February 29, 2024.

3. Leigh Collins, "First E-Fuel Made from Green Hydrogen and CO_2 Is 100 Times More Expensive Than Petrol, but Costs Should Plummet," *Hydrogen Insight*, March 27, 2023.

4. Jasper Jolly, "E-Fuels: How Big a Niche Can They Carve Out for Cars?," *The Guardian*, May 5, 2023.

5. Falko Ueckerdt et al., "Potential and Risks of Hydrogen-Based E-Fuels in Climate Change Mitigation," *Nature Climate Change*, May 2021.

6. Rose Oates, "Power-to-Liquids Primer: Fuel from Thin Air," BloombergNEF, May 2024.

7. Ghassan Wakim, "Hydrogen for Decarbonization: A Realistic Assessment," Clean Air Task Force, November 2023.

8. "NEOM Green Hydrogen Complex," Air Products website, accessed July 2024.

9. Adam Zewe, "Study Finds Health Risks in Switching Ships from Diesel to Ammonia Fuel," *MIT News*, July 11, 2024.

10. Colton Poore, "Ammonia Fuel Offers Great Benefits but Demands Careful Action," Princeton University, November 7, 2023.

11. Sam Meredith, "A Global Gold Rush for Buried Hydrogen Is Underway—as Hype Builds over Its Clean Energy Potential," CNBC, March 26, 2024.

12. Paul Martin, "Ammonia Pneumonia," LinkedIn, January 29, 2021, updated June 18, 2024.

13. Angeli Mehta, "In the Voyage to Net-Zero, Which Green Shipping Fuel Will Rule the Seas?," Reuters, May 15, 2023.

14. Alexander Budzier, "Five Strategies for Optimizing Power-to-X Projects," Boston Consulting Group in collaboration with Oxford Global Projects, September 28, 2023.

15. US Department of Energy, "Emerging Uses for Methanol," accessed August 2023.

16. International Renewable Energy Agency, "Innovation in Renewable Methanol," 2021.

17. Gustavo H. Ruffo, Autoevolution. "Synthetic Fuel Producers Betting on Methanol Must Have Forgotten It Is Toxic," March 18, 2024.

18. Maria Gallucci, "This Common Chemical Could Help Shipping Giants Start to Decarbonize," Canary Media, April 21, 2023.

19. Amira Nemmour, "Green Hydrogen-Based E-Fuels (E-Methane, E-Methanol, E-Ammonia) to Support Clean Energy Transition: A Literature Review," *International Journal of Hydrogen Energy*, September 2023.

20. Noa Leach, "New E-Fuels Could Help Save the Planet (and Your Car) but at a Hidden Cost," *BBC Science Focus*, June 1, 2023. The full quote is, "E-fuels are a good idea but we need to look at the consequences of their life cycle and the ability of our planet to sustainably supply the metals and materials needed." The reality is that e-fuels can't help save the planet or your car, and they aren't a "good idea" with or without caveats.

21. Karl Dums, "Forget Electric, Synthetic Fuels Can Save the World, Says Porsche." *CarExpert*, January 31, 2024.

22. Julian Spector, "How Hydrogen E-Fuels Can Power Big Ships and Planes," Canary Media, January 31, 2024.

23. Murray Douglas, "E-Fuels: Unlocking Hydrogen," Wood Mackenzie, June 2024.

24. John Ainger, "EU's Von Der Leyen Eyes E-Fuels Carve-Out for Cars, Canfin Says," Bloomberg, July 16, 2024.

25. Michael Barnard, "Air Carbon Capture's Scale Problem: 1.1 Astrodomes for a Ton of CO_2," CleanTechnica, March 14, 2019.

26. National Academy of Sciences, *Negative Emissions Technologies and Reliable Sequestration*, 2019. The efficiency being calculated here is the "exergy efficiency," or the so-called second law (of thermodynamics) efficiency. It is defined as "the ratio of minimum work to real work" (Wmin/Wreal).

27. June Sekera and Andreas Lichtenberger, "Assessing Carbon Capture: Public Policy, Science, and Societal Need," *Biophysical Economics and Sustainability*, October 6, 2020.

28. Noah McQueen et al., "Natural Gas vs. Electricity for Solvent-Based Direct Air Capture," *Frontiers in Climate*, January 27, 2021.

29. European Academies' Science Advisory Council (EASAC), "Forest Bioenergy Update: BECCS and Its Role in Integrated Assessment Models," February 2022. The EASAC consists of the twenty-seven national science academies of the EU Member States, Norway, Switzerland, and the United Kingdom. The report notes, "Policy-makers should suspend expectations that BECCS [bio-energy with carbon capture, utilization, and storage] can deliver significant [carbon dioxide] removals by 2050 until models have identified the sensitivity of atmospheric CO_2 levels to different feedstock payback times and can be confident that time-related targets can be achieved."

30. James Lattner and Scott Stevenson, "Renewable Power for Carbon Dioxide Mitigation," American Institute of Chemical Engineers, March 2021. The other significant advantage EVs have is that "DAC does not help balance variable supply with demand—the capacity of the DAC plant cannot be fully utilized on renewable supply alone."

31. Julian Spector. "EVs Can Be Cheaper on a Monthly Basis than Gas-Powered Cars," Canary Media, January 5, 2024.

32. Sudipta Chatterjee and Kuo-Wei Huang, "Unrealistic Energy and Materials Requirement for Direct Air Capture in Deep Mitigation Pathways," *Nature Communications*, July 2020.

33. June Sekera and Andreas Lichtenberger, "Assessing Carbon Capture: Public Policy, Science, and Societal Need," *Biophysical Economics and Sustainability*, October 2020.

34. Michael Cembalest, 2021 Annual Energy Paper, JP Morgan Asset and Wealth Management, 2021.

35. EIA, "Direct Air Capture 2022," April 2022.

36. Congressional Research Service (CRS), "Carbon Capture and Sequestration (CCS) in the United States," October 5, 2022, Washington, DC.

37. Adam Baylin-Stern and Niels Berghout, "Is Carbon Capture Too Expensive?," International Energy Agency, February 17, 2021.

38. Congressional Research Service (CRS), "Carbon Capture and Sequestration (CCS) in the United States," October 2021, Washington, DC.

39. Daniel Krekel et al., "The Separation of CO_2 from Ambient Air—A Techno-Economic Assessment," *Applied Energy*, May 2018.

40. Brian Kahn, "Removing Carbon from the Air Enters Its Awkward Teen Years," Bloomberg, June 12, 2023.

41. Microsoft, *Microsoft Carbon Removal: Lessons from an Early Corporate Purchase*, 2021. Most of the projects Microsoft selected were so-called natural climate solutions such as reforestation and improved forest management, which typically cost far less than $50 per ton of carbon removed.

42. Soheil Shayegh et al., "Future Prospects of Direct Air Capture Technologies, Insights from an Expert Elicitation Survey," *Frontiers in Climate*, May 2021.

43. Howard Herzog, "Chapter 6: Direct Air Capture," in Mai Bui and Niall Mac Dowell, eds., *Greenhouse Gas Removal Technologies*, Royal Society of Chemistry, August 2022.

44. John Young et al., "The Cost of Direct Air Capture and Storage Can Be Reduced via Strategic Deployment but Is Unlikely to Fall Below Stated Cost Targets," *One Earth*, July 21, 2023. The article notes, "Pairing [direct air capture and storage] to intermittent renewables is unlikely to be cost-effective unless the renewable power comes from installations with high-capacity factors, such as some offshore wind installations." A requirement for large amounts of high-capacity renewables would be a severe siting constraint.

45. Edward Rubin, "Improving Cost Estimates for Advanced Low-Carbon Power Plants," *International Journal of Greenhouse Gas Control*, September 2019.

46. Abhishek Malhotra and Tobias S. Schmidt, "Accelerating Low-Carbon Innovation," *Joule*, November 2020.

47. Dave Roberts, "Which Technologies Get Cheaper over Time, and Why? A Conversation with Abhishek Malhotra and Tobias Schmidt About Learning Curves," *Volts* podcast, January 13, 2023. The quote is from Malhotra.

48. S&P Global, "S&P Global Commodity Insights Releases Its Latest 2024 Energy Outlook," December 14, 2023.

49. Paul Martin, "Ammonia Pneumonia," LinkedIn, January 29, 2021, updated June 18, 2024.

50. IPCC, "Climate Change 2022: Mitigation of Climate Change," 2022.

51. IEA, "Energy Technology Perspectives 2017," June 2017.

52. IEA, "Net Zero Roadmap: A Global Pathway to Keep the 1.5°C Goal in Reach 2023 Update," September 2023.

53. Joseph Romm, "Why Scaling Bioenergy and Bioenergy with Carbon Capture and Storage (BECCS) Is Impractical and Would Speed Up Global Warming," University of Pennsylvania Center for Science, Sustainability and the Media," November 2023.

54. Sarah Hurtes and Weiyi Cai, "Europe Is Sacrificing Its Ancient Forests for Energy," *New York Times*, September 7, 2022.

55. Mary Booth, "Not Carbon Neutral: Assessing the Net Emissions Impact of Residues Burned for Bioenergy," *Environmental Research Letters*, February 2018.

56. Alessandro Agostini et al., "Carbon Accounting of Forest Bioenergy: Conclusions and Recommendations from a Critical Literature Review," Joint Research Centre (JRC) of the European Commission, 2014. See also Brian Kittler et al., "Environmental Implications of Increased Reliance of the

EU on Biomass from the South East US," European Commission, January 2016.

57. Michael Norton et al., "Serious Mismatches Continue Between Science and Policy in Forest Bioenergy," *GCB Bioenergy*, August 22, 2019. The article notes, "These conclusions on the limited amounts of residues available are consistent with monitoring by environmental groups which have tracked areas of clear-cut forests to pellet mills." See, for instance, Natural Resources Defense Council, "European Imports of Wood Pellets for 'Green Energy' Devastating US Forests," 2017, and Natural Resources Defense Council, "Global Markets for Biomass Energy are Devastating U.S. Forests," September 2022.

58. Mary S. Booth et al., "Not Carbon Neutral: Assessing the Net Emissions Impact of Residues Burned for Bioenergy," *Environmental Research Letters*, February 2018.

59. IPCC, "Climate Change 2022: Mitigation of Climate Change, Summary for Policymakers," Working Group III contribution to the Sixth Assessment Report, 2022.

60. John Sterman et al., "Does Wood Bioenergy Help or Harm the Climate?," *Bulletin of the Atomic Scientists*, May 2022. The study notes, "The younger the forest and faster it is growing when harvested for bioenergy, the more future carbon sequestration is lost."

61. Peter Raven et al., "Letter Regarding Use of Forests for Bioenergy," to the heads of the United States, European Union, Japan, and South Korea, February 11, 2021. See also Timothy D. Searchinger et al., "Europe's Renewable Energy Directive Poised to Harm Global Forests," *Nature Communications*, September 2018.

62. National Academies of Sciences, *Negative Emissions Technologies and Reliable Sequestration: A Research Agenda*, The National Academies Press, 2019. The Academies note that even in the case of wood residues there can be a long-lasting carbon debt: "If the wood residues would otherwise have been used in a long-lived product such as particle board, it could take decades for the use of this material for bioenergy to have a positive effect of reducing atmospheric CO_2."

63. Lydia J. Smith and Margaret S. Torn, "Ecological Limits to Terrestrial Biological Carbon Dioxide Removal," *Climatic Change*, February 2013.

64. Howard Herzog et al., "Comment on 'Reassessing the Efficiency Penalty from Carbon Capture in Coal-Fired Power Plants,'" *Environmental Science and Technology*, May 12, 2016.

65. Daniel Quiggin, "BECCS Deployment. The Risks of Policies Forging Ahead of the Evidence," Chatham House, London, UK, October 2021.

66. Bruce Robertson and Milad Mousavian, "The Carbon Capture Crux: Lessons Learned," Institute for Energy Economics and Financial Analysis, September 1, 2022.

67. John D. Sterman et al., "Does Replacing Coal with Wood Lower CO_2 Emissions, Dynamic Lifecycle Analysis of Wood Bioenergy," *Environmental Research Letters*, January 18, 2018. Sterman was part of the team responsible for the analysis presented here using the En-ROADS system dynamics model.

68. Roel Hammerschlag, "Uncaptured Biogenic Emissions of BECCS Fueled by Forestry Feedstocks," prepared for Natural Resources Defense Council, September 2021.

69. Natural Resources Defense Council, "A Bad Biomass Bet: Why the Leading Approach to Biomass Energy with Carbon Capture and Storage Isn't Carbon Negative," October 2021.

70. Climate Interactive, "The En-ROADS Climate Solutions Simulator," accessed November 2023.

71. Climate Interactive, "En-ROADS June 2023: Bioenergy," June 1, 2023.

72. European Academies' Science Advisory Council (EASAC), "Forest Bioenergy Update: BECCS and Its Role in Integrated Assessment Models," February 2022. See also "About EASAC," on the EASAC website. The EASAC consists of the twenty-seven national science academies of the EU Member States, Norway, Switzerland, and the United Kingdom.

73. Eric F. Lambin et al., "Global Land Use Change, Economic Globalization, and the Looming Land Scarcity," *Proceedings of the National Academy of Sciences*, February 2011.

74. Tim Searchinger et al., "Does the World Have Low-Carbon Bioenergy Potential from the Dedicated Use of Land?," *Energy Policy*, November 2017.

75. Peter Potapov et al., "Global Maps of Cropland Extent and Change Show Accelerated Cropland Expansion in the Twenty-First Century," *Nature Food*, 2022.

76. Kate Dooley et al., "The Land Gap Report," November 2022.

77. Joseph Romm, "The Next Dust Bowl," *Nature*, October 2011.

78. UN Food and Agriculture Organization, *The State of the World's Land and Water Resources for Food and Agriculture: Systems at Breaking Point*, 2021.

79. Bojana Bajželj et al., "Importance of Food-Demand Management for Climate Mitigation," *Nature Climate Change*, August 2014.

80. David Fickling, "It's Time to Get Biofuels out of Your Gas Tank Analysis," Bloomberg, June 9, 2022.

81. IEA, "The Role of E-fuels in Decarbonising Transport," December 2023.

82. Timothy D. Searchinger, "Do Biofuel Policies Seek to Cut Emissions by Cutting Food?," *Science*, March 2015.

83. Melissa J. Scully, "Carbon Intensity of Corn Ethanol in the United States," *Environmental Research Letters*, March 2021.

84. Tim Searchinger and Ralph Heimlich, "Avoiding Bioenergy Competition for Food Crops and Land," World Resources Institute, January 2015.

85. Tyler J. Lark et al., "Environmental Outcomes of the US Renewable Fuel Standard," *Proceedings of the National Academy of Sciences*, February 2022. The study also looked at other lifecycle analyses of corn ethanol GHG emissions, including the Environmental Protection Agency's, and found that "incorporating the domestic LUC [land use change] emissions from our analysis into other fuel program estimates similarly annuls or reverses the GHG advantages they calculate for ethanol relative to gasoline."

86. Mark Z. Jacobson, "Should Transportation Be Transitioned to Ethanol with Carbon Capture and Pipelines or Electricity? A Case Study," *Environmental Science & Technology*, October 26, 2023.

87. Amena H. Saiyid, "Biofuels Offer Low-Hanging Fruit in Nascent Carbon Capture Industry," Cipher, March 13, 2024.

88. Biodiesel made from palm oil is more problematic than bioethanol. It requires much more land per mile driven and has caused considerable destruction of carbon-rich peatland forests in Southeast Asia. The *New York Times* wrote in 2018 that "six of the world's leading carbon-modeling schemes, including the E.P.A.'s, have concluded that biodiesel made from Indonesian palm oil makes the global carbon problem worse, not better." See Abrahm Lustgarten, "Palm Oil Was Supposed to Help Save the Planet. Instead It Unleashed a Catastrophe," *New York Times*, November 20, 2018.

89. H. K. Gibbs et al., "Tropical Forests Were the Primary Sources of New Agricultural Land in the 1980s and 1990s," *Proceedings of the National Academy of Sciences*, August 2010.

90. Gabriel Popkin, "Cropland Has Gobbled Up over 1 Million Square Kilometers of Earth's Surface," *Science*, December 2021. See also Peter Potapov et al., "Global Maps of Cropland Extent and Change Show Accelerated Cropland Expansion in the Twenty-First Century," *Nature Food*, December 2021.

91. Tyler J. Lark et al., "Cropland Expansion in the United States Produces Marginal Yields at High Costs to Wildlife," *Nature Communications*, September 2020.

92. Tyler J. Lark et al. (2020) noted that "the large area of long-term grassland conversion we identified provides strong evidence of widespread clearing of previously undisturbed lands."

93. Seth A Spawn et al., "Carbon Emissions from Cropland Expansion in the United States," *Environmental Research Letters*, April 2019. See also Zhen Yu

et al., "Largely Underestimated Carbon Emission from Land Use and Land Cover [LULAC] Change in the Conterminous US," *Global Change Biology*, July 2019.

94. UN Food and Agriculture Organization (FAO), "Global Forest Resources Assessment 2020 Remote Sensing Survey," May 2022. See also FAO, "FAO Remote Sensing Survey Reveals," 2022.

95. Pawlok Dass et al., "Grasslands May Be More Reliable Carbon Sinks than Forests in California, *Environmental Research Letters*, July 2018.

96. César Terrer et al., "A Trade-Off Between Plant and Soil Carbon Storage Under Elevated CO_2," *Nature*, March 2021.

97. Stanford University news release, "Soils or Plants Will Absorb More CO_2 as Carbon Levels Rise—but Not Both, Stanford Study Finds," March 24, 2021.

98. National Academy of Sciences, *Negative Emissions Technologies and Reliable Sequestration: A Research Agenda*, The National Academies Press, 2019.

99. Felix Creutzig et al., "Considering Sustainability Thresholds for BECCS in IPCC and Biodiversity Assessments," *GCB Bioenergy*, February 2021. See also Anthony Young, "Is There Really Spare Land? A Critique of Estimates of Available Cultivable Land in Developing Countries," *Environment, Development and Sustainability*, March 1999.

100. Kate Dooley et al., "The Land Gap Report," November 2022.

101. Walter V. Reid et al., "The Future of Bioenergy," *Global Change Biology*, January 2020.

102. Tim Searchinger, "Global Consequences of the Bioenergy Greenhouse Gas Accounting Error," in Oliver Inderwildi and Sir David King (eds.), *Energy, Transport, & the Environment*, Springer, 2012.

103. Walter V. Reid et al., "The Future of Bioenergy," *Global Change Biology*, January 2020.

104. Material Economics, "EU Biomass Use in a Net-Zero Economy: A Course Correction for EU Biomass," 2021.

Chapter 7

1. Dan McCarthy, "Chart: One in Five New Cars Sold This Year Will Be Battery-Powered," Canary Media, May 3, 2024.

2. Global EV, "Trends in Electric Cars," International Energy Agency, 2024.

3. Leigh Collins, "Global Sales of Hydrogen Vehicles Fell by More Than 30% Last Year, with China Becoming World's Largest Market," *Hydrogen Insight*, January 5, 2024.

4. Mark Kane, "US Hydrogen Car Sales Q2 2024," InsideEVs, August 5, 2024.

5. Phil Wrigglesworth, "The Car of the Perpetual Future," *The Economist*, September 6, 2008.

6. Peter Flynn, "Commercializing an Alternate Vehicle Fuel: Lessons Learned from Natural Gas for Vehicles," *Energy Policy*, June 2002.

7. Department of Transportation, "Fuel Options for Reducing Greenhouse Gas Emissions from Motor Vehicles," 2003.

8. Leigh Collins, "Global Sales of Hydrogen Vehicles Fell by More Than 30% Last Year, with China Becoming World's Largest Market," Hydrogen Insight, January 5, 2024.

9. SNE Research, "In Q1 2024, Global FCEV Market with a 36.4% YoY Degrowth," May 13, 2024.

10. Ford Motor Company, *Sustainability Report 2013/14*, June 18, 2014.

11. Polly Martin, "Toyota's New Hydrogen Car to Be Significantly More Expensive Than Its Entry-Level H2-Powered Mirai," *Hydrogen Insight*, November 6, 2023.

12. Leigh Collins, "Analysis: It Is Now Almost 14 Times More Expensive to Drive a Toyota Hydrogen Car in California Than a Comparable Tesla EV," *Hydrogen Insight*, January 5, 2024.

13. Todd Woody, "California's Hydrogen Fuel Cell Cars Lose Traction Against Battery Models," Bloomberg, April 4, 2024.

14. Miki Crowell and Andrew Martinez. *2023 Annual Assessment of the Hydrogen Refueling Network in California*. California Energy Commission and California Air Resources Board, December 2023. Publication Number: CEC-600-2023-069.

15. Barry C. Lynn, "Hydrogen's Dirty Secret," *Mother Jones*, May–June 2003.

16. Felix Salmon, "Gas Stations Closing as Prices Rise," *Axios*, July 15, 2022. There are many different estimates for that number.

17. Ulf Bossel and Baldur Eliasson, "*Energy and the Hydrogen Economy*," January 2003, Raymond Drnevich, personal communication. Drnevich says, "If you are looking for the energy needed to compress hydrogen for storage on board a vehicle, it is necessary to consider the overpressure required to fill the vehicle to the desired end pressure in a reasonable period of time." For 5,000 psi onboard storage, this might require an overpressure of 7,000 psi.

18. George Thomas and Jay Keller, "Hydrogen Storage—Overview," presentation to the H_2 Delivery and Infrastructure Workshop, Washington, DC, May 7–8, 2003. It had been said that the hydrogen tank volume might only have to be four times the volume for a tank with a comparable driving range, because the fuel cell vehicle was considered to be twice as fuel efficient.

19. Dale Simbeck and Elaine Chang, "Hydrogen Supply: Cost Estimate for Hydrogen Pathways—Scoping Analysis," Department of Energy, National Renewable Energy Laboratory, July 2002.

20. National Academies of Sciences, "The Hydrogen Economy: Opportunities, Costs, Barriers, and R&D Needs," 2004.

21. American Physical Society, "The Hydrogen Initiative," March 2004.

22. JoAnn Milliken et al., "Hydrogen Storage Activities Under the Freedom Car & Fuel Initiative," presentation to the National Hydrogen Association's Fourteenth Annual US Hydrogen Conference and Hydrogen Expo, Washington, DC, March 4–6, 2003.

23. Mariya Koleva and Marc Melaina, "Hydrogen Fueling Station Cost," US Department of Energy, February 11, 2020.

24. US Department of Energy, "Alternative Fueling Station Locator," accessed August 2024.

25. US Department of Energy, "Liquid Hydrogen Delivery," accessed August 2024.

26. Hydrogen Fuel Cell Partnership, "The 1st Hydrogen Bank: Achieving Zero-Boil-Off Liquid Hydrogen Storage," May 18, 2022.

27. Companies and governments are pursuing technology to minimize boil-off. But bringing it down to acceptably low levels will require new technology. In fact, in 2024 the DOE made grants totaling $18 million for new designs to do that. But even if such solutions are developed and commercialized in future years, they will still require a costly replacement of a large fraction of the current liquid hydrogen storage and distribution infrastructure, including all of the tanker trucks. Who is going to pay for all that?

28. Ulf Bossel and Baldur Eliasson, "Energy and the Hydrogen Economy," January 2003. The canister trucks carry 500 kilograms of hydrogen but deliver only 400 kilograms. These canisters typically don't discharge all their fuel because extra compressors would be needed to completely empty the delivery tank (once the pressure in the delivery tank dropped below the pressure in the receiving tank), complicating the process and necessitating more capital equipment at the fueling station.

29. Michael Liebreich, X.com, November 27, 2022.

30. Marianne Mintz et al., "Hydrogen: On the Horizon or Just a Mirage?," *SAE Transactions: Journal of Fuels and Lubricants*, 2002.

31. American Petroleum Institute, "Where Are the Pipelines?," accessed August 2024.

32. Electric Power Research Institute, "Regional Hydrogen Pipeline Costs for US-REGEN Model," April 30, 2024.

33. Mathias Zacarias and Jane Nakano, "Exploring Hydrogen Midstream

Distribution and Delivery," Center for Strategic and International Studies, April 28, 2023.

34. National Academies of Sciences, "The Hydrogen Economy: Opportunities, Costs, Barriers, and R&D Needs," 2004.

35. Kyle Hyatt, "Electric Car Fires: What You Should Know," Edmunds.com, March 5, 2024. See also Australian Government, "EV Fire FAQs," at EV Fire Safe, July 2021.

36. Liejin Guo et al., Hydrogen Safety: An Obstacle That Must Be Overcome on the Road Towards Future Hydrogen Economy," *International Journal of Hydrogen Energy*, January 2, 2024.

37. Arthur D. Little Inc., "Guidance for Transportation Technologies: Fuel Choice for Fuel-Cell Vehicles," final report to DOE, February 6, 2002.

38. James Hansel (Air Products and Chemicals Inc.), "Safety Considerations for Handling Hydrogen," presentation to the Ford Motor Company, Allentown, PA, June 12, 1998.

39. Jim Campbell (Air Liquide), "Hydrogen Delivery Technologies and Systems: Pipeline Transmission of Hydrogen," presentation to the Strategic Initiatives for Hydrogen Delivery Workshop, Washington, DC, May 7–8, 2003.

40. James Hansel, "Safety Considerations for Handling Hydrogen," presentation to the Ford Motor Company, Allentown, PA, June 12, 1998.

41. Paul M. Ordin, "Safety Standard for Hydrogen and Hydrogen Systems," NASA Office of Safety and Mission Assurance, February 12, 1997.

42. The 22% figure comes from Russell Moy, "Tort Law Considerations for the Hydrogen Economy," *Energy Law Journal*, November 2003. The 40% figure comes from James Hansel, "Safety Considerations for Handling Hydrogen" presentation to the Ford Motor Company, June 12, 1998.

43. Russell Moy, "Tort Law Considerations for the Hydrogen Economy," *Energy Law* Journal, November 2003. See also Russell Moy, "Liability and the Hydrogen Economy," *Science*, July 4, 2003.

44. Jim Campbell (Air Liquide), "Hydrogen Delivery Technologies and Systems: Pipeline Transmission of Hydrogen," presentation to the Strategic Initiatives for Hydrogen Delivery Workshop, Washington, DC, May 7–8, 2003.

45. Dan Sperling and Joan Ogden, "The Hope for Hydrogen," *Issues in Science and Technology*, Spring 2004.

46. C. A. Geffen et al., "Transportation and Climate Change: The Potential for Hydrogen Systems," SAE Technical Paper 2004-01-0700, 2004.

47. Robert Edwards et al., "Well-to-Wheels Analysis of Future Automotive Fuels and Powertrains in the European Context," European Commission Center for Joint Research, January 2004.

48. Ulf Bossel, "The Hydrogen 'Illusion': Why Electrons Are a Better Energy Carrier," *Cogeneration & On-Site Power Production*, March–April 2004.
49. Alec Brooks, "Carb's Fuel Cell Detour on the Road to Zero Emission Vehicles," EVnut.com, May 2, 2004.
50. Joe Romm, "Everything You Could Want to Know About the Plug-In Hybrid and Electric Vehicle Announcements," *ThinkProgress*, January 14, 2009.

Conclusion

1. NASA, "NASA Analysis Confirms 2023 as Warmest Year on Record," January 12, 2024.
2. Bella Isaacs-Thomas, "'We're Frankly Astonished.' Why 2023's Record-Breaking Heat Surprised Scientists," *PBS News*, January 19, 2024.
3. David T. Ho, "Carbon Dioxide Removal Is Not a Current Climate Solution— We Need to Change the Narrative," *Nature*, April 4, 2023.
4. Adam Vaughan, "World Has No Capacity to Absorb New Fossil Fuel Plants, Warns IEA," *The Guardian*, November 13, 2018.
5. Maria Sand et al., "A Multi-Model Assessment of the Global Warming Potential of Hydrogen," *Nature Communications Earth & Environment*, June 7, 2023.
6. Jasmin Cooper et al., "Hydrogen Emissions from the Hydrogen Value Chain—Emissions Profile and Impact to Global Warming," *Science of the Total Environment*, July 15, 2022. The study concludes, "It is likely that CO_2 and non-CO_2 GHG emissions in some, if not all, H_2 supply chains are more important." But the authors are using the old GWP100 of hydrogen rather than the new GWP20, which is seven times higher, so hydrogen emissions may well dominate the warming impact in a great many supply chains.
7. Zhiyuan Fan et al., "Hydrogen Leakage: A Potential Risk for the Hydrogen Economy," Columbia University Center on Global Energy Policy, July 2022.
8. Alessandro Arrigoni and Laura Bravo Diaz, "Hydrogen Emissions from a Hydrogen Economy and Their Potential Global Warming Impact," European Commission Joint Research Centre, August 18, 2022. The Air Liquide company has a design goal of 4% to 5% leakage for 2030. That seems optimistic but would, in any case, be at least ten times higher than a tolerable leakage rate from a climate perspective.
9. Paul Wolfram, "Using Ammonia as a Shipping Fuel Could Disturb the Nitrogen Cycle," *Nature Energy*, October 5, 2022.
10. Alexander Budzier et al., "Five Strategies for Optimizing Power-to-X Projects," Boston Consulting Group, September 28, 2023.
11. S&P Global Commodities, "S&P Global Commodity Insights Releases Its Latest 2024 Energy Outlook," December 14, 2023.

12. Xiaoting Wang, "Electrolysis System Cost Forecast 2050: Higher for Longer," BloombergNEF, September 12, 2024. See also Martin Tengler, head of hydrogen research at BloombergNEF, LinkedIn post, September 2024.
13. Rachel Morison and Petra Sorge, "Europe's Spending Billions on Green Hydrogen. It's a Risky Gamble," Bloomberg, May 17, 2024.
14. International Energy Agency, "Global Hydrogen Review 2023," September 2023.
15. Simon Bennett, personal communications, August 2024.
16. Prachi Patel, "Home Heating with Hydrogen: Ill-Advised as It Sounds," *IEEE Spectrum*, October 4, 2022.
17. American Council for an Energy-Efficient Economy, "Industrial Heat Pumps," accessed August 6, 2024.
18. Silvia Madeddu et al., "The CO_2 Reduction Potential for the European Industry via Direct Electrification of Heat Supply (Power-to-Heat)," *Environmental Research Letters*, November 2020.
19. Paul Martin, "Hydrogen Hype Is Crashing, but We Can't Afford to Give Up on Renewable Hydrogen," Hydrogen Science Coalition, August 1, 2024.
20. Diana Kinch, "Direct-Reduced Iron Becomes Steel Decarbonization Winner," *S&P Global*, June 22, 2022.
21. Agora Industry and Wuppertal Institute and Lund University, *Low-Carbon Technologies for the Global Steel Transformation*, 2024.
22. Simon Nicholas and Soroush Basirat, "New from Old: The Global Potential for More Scrap Steel Recycling," The Institute for Energy Economics and Financial Analysis, December 2021.
23. Angela Macdonald-Smith, Peter Ker, and Jessica Sier, "Green Hydrogen Too 'Expensive and Inefficient': Finkel," *Financial Review*, July 18, 2024.
24. Ghassan Wakim et al., "Hydrogen in the Power Sector: Limited Prospects in a Decarbonized Electric Grid," Clean Air Task Force, June 2024.
25. Clean Air Task Force (CATF), "Dedicated Clean Hydrogen Production Likely Has a Limited Role in Power Sector Decarbonization, Finds CATF Report," CATF News Release, July 4, 2024.
26. Rose Oates, "Power-to-Liquids Primer: Fuel from Thin Air," BloombergNEF, May 2024.
27. Polly Martin, "Why Shipping Is Opting for Green Hydrogen-Based Methanol over Ammonia Despite Much Higher Fuel Costs," *Hydrogen Insight*, January 9, 2024.
28. "China Launches First 700-TEU Electric Containership for Yangtze Service," *The Maritime Executive*, July 28, 2023.

About the Author

Dr. Joseph Romm was principal deputy and then acting assistant secretary of energy efficiency and renewable energy in the mid-1990s, overseeing a $1 billion budget on climate solutions for transportation, buildings, and industry, including hydrogen, fuel cells, energy storage, bioenergy, and renewables. He is a senior research fellow at the University of Pennsylvania Center for Science, Sustainability, and the Media, where he researches climate solutions. In 2009, *Rolling Stone* named him one of 100 "People Who Are Reinventing America," and *Time* named him "Hero of the Environment." He received the 2024 Ban Ki-Moon Award for Environmental Leadership from the former UN Secretary General. He has authored eleven books, including the original *The Hype About Hydrogen*, which was named one of the best science and technology books of 2004 by *Library Journal*. His TEDx talk is "The Surprising Truth About Solving Climate Change." He holds a PhD in physics from MIT.